Innovations in Process Engineering 2022

Held at the 2022 AIChE Annual Meeting

Phoenix, Arizona, USA
13-18 November 2022

ISBN: 978-1-7138-7892-6

Printed from e-media with permission by:

Curran Associates, Inc.
57 Morehouse Lane
Red Hook, NY 12571

Some format issues inherent in the e-media version may also appear in this print version.

AIChE
120 Wall Street, FL 23
New York, NY 10005-4020

Phone: (800) 242-4363
Fax: (203) 775-5177

www.aiche.org

Additional copies of this publication are available from:

Curran Associates, Inc.
57 Morehouse Lane
Red Hook, NY 12571 USA
Phone: 845-758-0400
Fax: 845-758-2633
Email: curran@proceedings.com
Web: www.proceedings.com

TABLE OF CONTENTS

EFFICIENT PROCESSING OF LIGNIN TO BIOPRODUCTS AND BIOFUELS

INTEGRATED PROCESS ENGINEERING AND ECONOMIC ANALYSIS

IONIC LIQUIDS: NOVEL SEPARATION, CATALYTIC REACTION AND ELECTROCHEMICAL PROCESSES

MATERIALS FOR THERMOCHEMICAL AND ELECTROCHEMICAL ENERGY STORAGE

TOPICAL PLENARY: FRONTIERS IN GREEN PROCESS AND PRODUCT ENGINEERING (INVITED TALKS)

PROCESS INTENSIFICATION AND MODULAR MANUFACTURING: INTENSIFIED REACTION AND SEPARATION PROCESSES

320a Automated, Multi-Task Bayesian Optimization of Pharmaceutical Processes

Connor Taylor

In recent years, there has been an increased interest in the use of automated, self-optimizing continuous flow platforms for the optimization of chemical processes.[1] This has arisen from the desire for faster reaction optimization, improved process sustainability and cheaper overall process development. The use of these platforms aims to enhance the capabilities of the researcher by removing the need for labour-intensive experimentation, allowing them to focus on more challenging tasks. These platforms use optimization (machine-learning) algorithms to use only minimal reaction material, but ultimately gain the most process information possible, making their deployment in fine chemical and pharmaceutical industries very attractive.[2]

Recent work has shown that Bayesian optimization in particular is a powerful tool for these applications.[3] However, in all previous studies, each optimization has begun using no *a priori* information about the chemical landscape for the reaction of interest. This protocol therefore requires many initial experimental iterations whereby the algorithm is learning about the experimental design space, without any information on where the optimal reaction conditions may be. This can be wasteful in terms of both cost and time, particularly when there may already be a wealth of information on the chemical transformation of interest from previous optimization campaigns. For medicinal chemistry applications, such as developing chemistries for the elaboration of small polar fragments into nanomolar leads (i.e. Fragment-Based Drug Discovery)[4,5], the use of efficient optimization techniques is paramount due to the minimal quantity and increased preciousness of starting materials.

This work shows the first example of leveraging previous reaction optimization data for unseen chemical transformations, using multi-task Bayesian optimization (MTBO). The framework of MTBO, first introduced by Swerksy *et al.*,[6] replaces the standard probabilistic model in Bayesian optimization with a multi-task model. As these multi-task models can be trained on data from related tasks, we can therefore utilise data from similar reactions - both from the laboratory and from the literature. In this work, we sequentially optimize the reaction conditions for five similar palladium-catalyzed C-H activation reactions using our automated reactor platform. Each reaction yields a pharmaceutically relevant oxindole product, as the optimization efficiency increases with further experimental data - this is true even with different chemical substrates that host different functionalities. One of these five processes is shown in the corresponding scheme, highlighting the reactivity that we targeted in this study.

This work represents a step-change in self-optimization, enabling ultra-efficient reaction optimization through the use of previous experimental data.

References: [1] *React. Chem. Eng.*, 2019, 4, 1545-1554. [2] *Chem. Eng. J.*, 2020, 384, 123340. [3] *Nature*, 2021, 590, 89-96. [4] *Nat. Chem.*, 2009, 1, 187-192. [5] *Chem. Sci.*, 2021, 12, 11976-11985. [6] *NIPS Proc.*, 2013, 2004-2012.

320b Chemical Looping Hydrogenation with Metal Oxide Bronzes for Selective Hydrogen Activation and Utilization

Yifan Deng, Evan Miu, Giannis Mpourmpakis, James R. McKone, Goetz Veser

Hydrogen is widely used in the chemical industry for hydrogenation reactions, including hydrotreating, ammonia synthesis, methanol synthesis, and a wide range of related processes. However, industrial hydrogenation reactions require high-purity H_2 streams. As hydrogen is currently produced almost exclusively via steam reforming of methane, which produces a mix of hydrogen and CO (syngas), production of high-purity hydrogen requires energy-intensive purification processes like water-gas shift and pressure-swing adsorption. Reduction of the energy demand and cost associated with these steps could have significant impact on future hydrogen utilization.

We are presenting a novel approach towards using low-purity hydrogen streams for thermocatalytic hydrogenation reactions via chemical looping hydrogenation (CLH). Chemical looping has been studied extensively for fuel combustion and reforming reactions where it serves to selectively transport oxygen from air to a fuel oxidation step, allowing partial or total oxidation of fuels without the need for (external) air separation and without diluting the product stream with N_2. Prior work on chemical looping has been almost exclusively used for oxygen transport and, to a lesser degree, activating nitrogen and carbon. Our work constitutes the first extension of the "looping principle" to the transport and purification of hydrogen.

Our approach uses catalyst-decorated metal oxide hydrogen bronzes as H carriers and the partial hydrogenation of acetylene (PHA) to ethylene as a test reaction. Using density functional theory and electroanalytical measurements, we have identified Pt-loaded tungsten trioxide, Pt-WO_3, as a suitable H carrier. On this carrier, H_2 is first activated through dissociation at Pt sites, and then inserted into the WO_3 via hydrogen spillover/intercalation. The carrier is then exposed to a pure acetylene stream in which WO_3 release H to hydrogenate C_2H_2 with high selectivity towards ethylene (>85%). The hydrogenation selectivity could be further controlled by tuning the degree of H intercalation into WO_3. Most significantly, the H loading of the carrier could be conducted directly with a syngas mixture containing as little as 15 mol% H_2 without loss of activity or selectivity toward PHA, demonstrating the ability of the proposed process to use impure H_2 streams directly for catalytic hydrogenation reactions without the need for costly and energy-intensive H_2 separation and purification steps.

320c General Techno-Economic Analysis of Electrochemical Ammonia Production Processes with Various Nitrogen Sources

Juyeon Kim, Woong Choi, Suhyun Lee, Jonggeol Na, Yun Jeong Hwang

Ammonia (NH_3) is one of the most important industrial chemicals, which has been widely used in agricultural and industrial fields as a feedstock for fertilizer, refrigerant, explosives, etc.[1] The global NH_3 market in 2020 was US$63.7 billion per year, and it is expected to reach US$ 89.6 billion per year by growing at 6.9% of the compound annual growth rate.[2] Also, the NH_3 composed of nitrogen and hydrogen could be a promising hydrogen carrier without concerns of carbon dioxide emission, which incorporates it as a promising energy storage material.[3]

The NH_3 has been commercially produced by Haber-Bosch (HB) process reacting nitrogen and hydrogen at high temperature (400–500°C) and high pressure (20–30 MPa) over Fe-based catalysts.[4] Although the HB process enables large-scale ammonia production, a large amount of energy and resource is required for the harsh operating conditions. Consequently, the global HB process consumed more than 1% of global energy demands, which contributed to a significant portion of greenhouse gas (GHG) emissions.[5]

To overcome this limitation, we focused on developing an electrochemical NH_3 production process based on renewable power. The electrochemical NH_3 production is based on the reduction of nitrogen sources (N_2, NO_x, etc.) obtained from air or wastewater, and it could be free from GHG emission and resources depletion by using renewable electricity.[6] Although previous experimental studies reported the potential of electrochemical NH_3 production via nitrogen reduction reaction (NRR)[6] or nitrate reduction reaction (NO_3RR)[7], few studies have been conducted on the system-level of techno-economic analysis and life cycle assessment for the electrochemical NH_3 production process.[8] In this study, we analyzed the electrochemical NH_3 production processes as a potential alternative to the HB process based on the techno-economic feasibility and environmental impact. Furthermore, we present a one-step system for the production of NH_3 from NO_3 in a cascade structure in which reaction, N_2 is oxidized to NO_3 (N_2OR) at the anode of the electrolyzer, and the NO_3 is supplied to the cathode to produce NH_3. Here, we focus on the conceptual process designs based on three electrocatalytic Nitrogen reductions reactions (NRR and NO_3RR, and N_2OR) in the NH_3 production process. First, we proposed an integrated system for the production of NH_3 based on three types of nitrogen reduction, which consist of nitrogen-source reduction (electrolyzer) and NH_3 separation (flash and distillation columns). Second, a techno-economic analysis was conducted on the three proposed processes. An automated process simulation and evaluation framework were developed to evaluate the impacts of various parameters of electrochemical reduction on the economics of electrochemical NH_3 production.

The automated framework of techno-economic analysis with rigorous process simulation managed by the central platform (Matlab) was employed to perform extensive sensitivity analysis. The process simulation is automatically performed by running commands in the Matlab code, and the Aspen Plus simulation results are transmitted back to the techno-economic evaluation code to determine the MSP of NH_3 of each process. Through global sensitivity analysis, the importance of significant parameters of electrochemical reactions such as current density, Faraday efficiency, and overpotential was analyzed to evaluate the performance of the design variables of the electrochemical process. The analysis was also conducted on several scenarios in which electrochemically produced NH_3 can compete in price with NH_3 of the existing HB process. To evaluate the environmental impact of the electrochemical ammonia production process, cradle-to-gate life cycle assessments for the electrochemical process and HB process were performed using SimaPro software. The two systems using NO3RR show the price of \$ 4-6 /kg at 2000 mA/cm^2, 1 V overpotential, and 90% of Faray efficiency, which suggests that it is possible to compete with the HB process if the power generation cost of renewable energy and the installation cost of the electrolyzer system is reduced through technological improvement in the future. The minimum selling price of the one-step system (N_2-NO_3-NH_3) was \$ 2.5/kg, which showed the lowest price compared to the N_2RR and NO_3RR systems in the future-based scenario. In this study, it is crucial that the environmental impact of the electrochemical process are significantly dependent on the carbon intensity of the electricity used. The operation of an electrochemical system with renewable energy would result in low climate change impact than HB process.

Finally, we proposed the bottlenecks of electrochemical NH_3 production and a feasible range of electrochemical reduction performance to replace the HB process. We hope that the present study will guide the electrochemical ammonia production to target overcoming the technological goals to make electrochemical reduction economically and environmentally compelling compared to the HB process.

1. MacFarlane, D. R.; Cherepanov, P. V.; Choi, J.; Suryanto, B. H.; Hodgetts, R. Y.; Bakker, J. M.; Vallana, F. M. F.; Simonov, A. N., A roadmap to the ammonia economy. *Joule* **2020,** *4* (6), 1186-1205.
2. *Global Ammonia Industry: Global Industry Analysts; 2022*; 2020.
3. Wang, G.; Mitsos, A.; Marquardt, W., Conceptual design of ammonia-based energy storage system: System design and time-invariant performance. *AIChE Journal* **2017,** *63* (5), 1620-1637.
4. Kandemir, T.; Schuster, M. E.; Senyshyn, A.; Behrens, M.; Schlögl, R., The Haber–Bosch process revisited: On the real structure and stability of "ammonia iron" under working conditions. *Angewandte Chemie International Edition* **2013,** *52* (48), 12723-12726.
5. Smith, C.; Hill, A. K.; Torrente-Murciano, L., Current and future role of Haber–Bosch ammonia in a carbon-free energy landscape. *Energy andamp; Environmental Science* **2020,** *13* (2), 331-344.
6. Soloveichik, G., Electrochemical synthesis of ammonia as a potential alternative to the Haber–Bosch process. *Nature Catalysis* **2019,** *2* (5), 377-380.
7. Wu, Z.-Y.; Karamad, M.; Yong, X.; Huang, Q.; Cullen, D. A.; Zhu, P.; Xia, C.; Xiao, Q.; Shakouri, M.; Chen, F.-Y., Electrochemical ammonia synthesis via nitrate reduction on Fe single atom catalyst. *Nature communications* **2021,** *12* (1), 1-10.
8. Gomez, J. R.; Baca, J.; Garzon, F., Techno-economic analysis and life cycle assessment for electrochemical ammonia production using proton conducting membrane. *International Journal of Hydrogen Energy* **2020,** *45* (1), 721-737.

320d Fractionation of Lignocellulosic Components By Environmental Benign Solvents

Zhimin Xue

The utilization of lignocellulosic components to materials, energy and chemicals are negative carbon processes. Fractionation of lignocellulosic components to lignin, cellulose, and semi-cellulose is a prerequisite for efficiently transformation of lignocellulosic components to valuable chemicals or energy. However, the components of lignocellulose are bonded together to form Lignin–carbohydrate complexes. It is hard to fractionation them to get pure components. We developed several environmental benign solvents to achieve the above goal.[1-5]

γ-Valerolactone (GVL)-based solvent[1], deep eutectic solvents (DESs) with different functional groups[2] and superbase-based DESs[3] for dissolution of lignins. Several solvent systems, including biomass-derived GVL and a second component of water, ionic liquids (ILs), DMSO, or DMF; DESs employing lactic acid and N-methylthiourea as HBDs and four different quaternary ammonium salts as HBAs; 17 kinds of 1,5-diazabicyclo[5.4.0]-5-undecene-based DESs, are designed for dissolution of various industrial lignins including dealkaline lignin (DAL), enzymatic hydrolysis lignin (EHL), sodium lignosulfonate (SL), and organosolv lignin (OL). The solubility of lignin in GVL-based solvent and hydrogen bond basicity parameter β value of solvents depends on both solvent and lignin, and the dissolved lignin can be recovered by ethanol. The functional groups in DESs significantly affected their ability to dissolve lignin, and a positive correlation was found between the solubility of lignin and the subtraction value of Kamlet–Taft empirical parameters α and β for the dissolution of EHL, DAL and OL. The solubility of lignin in superbase-based DESs significantly depended on the types of the DESs and lignin. Also the relationship between lignin solubility and the solvatochromic parameters α, β, and π* varied significantly based on the DESs and the types of lignin.

Several IL-based DESs using four ILs as the HBAs and four different HBDs, including thiourea (TU), N-methylthiourea (NMTU), glycerol (Gly), and ethylene glycol (EG) were designed and found as efficient solvents for xylan.[4] The xylan solubility in DES 1-allyl-3-methylimidazolium chloride plus EG could reach 40.4 wt% at 343.15 K, which is much higher than that in the corresponding IL. The enhancement effect might be ascribed to that the decrease of viscosity and increase the hydrogen bond accepting ability. Additionally, the regeneration of the xylan from the DESs could be easily realized using ethanol as the anti-solvent, and the regenerated xylan kept a high consistency with the original xylan.

Six kinds of novel superbase-based solvents were designed by 1,5-diazabicyclo[4.3.0]non-5-ene (DBN) with pyridine *N*-oxide (PyO) or 2-picoline-*N*-oxide (PiO) for dissolution of cellulose.[5] The solvents (denoted as DBN-PyO-x and DBN-PiO-4) could efficiently dissolve cellulose at mild temperatures (<80 ◦C). DBN-PyO-4 has a cellulose solubility of 14.1 wt% at 70 ◦C.

More robust solvents both environmental benign and cheap are anticipated to develop for industrial fractionation of lignocellulosic components. Moreover, the dissolution mechanism are far from disclosed. More efforts have to be devoted to this topic for the industrialization.

References:

[1] Xue, Z., Zhao, X., Sun, R. C., andamp; Mu, T. (2016). Biomass-derived γ-valerolactone-based solvent systems for highly efficient dissolution of various lignins: Dissolution behavior and

mechanism study. *ACS Sustainable Chemistry andamp; Engineering*, *4*(7), 3864-3870.

[2] Liu, Q., Zhao, X., Yu, D., Yu, H., Zhang, Y., Xue, Z., andamp; Mu, T. (2019). Novel deep eutectic solvents with different functional groups towards highly efficient dissolution of lignin. *Green Chemistry*, *21*(19), 5291-5297.

[3] Liu, Q., Mou, H., Chen, W., Zhao, X., Yu, H., Xue, Z., andamp; Mu, T. (2019). Highly efficient dissolution of lignin by eutectic molecular liquids. *Industrial andamp; Engineering Chemistry Research*, *58*(51), 23438-23444.

[4] Yu, H., Xue, Z., Lan, X., Liu, Q., Shi, R., andamp; Mu, T. (2020). Highly efficient dissolution of xylan in ionic liquid-based deep eutectic solvents. *Cellulose*, *27*(11), 6175-6188.

[5] Liu, Q., Yu, H., Mu, T., Xue, Z., andamp; Xu, F. (2021). Robust superbase-based emerging solvents for highly efficient dissolution of cellulose. *Carbohydrate Polymers*, *272*, 118454.

320e Construction of an Enzymatic Shuttling Compartment Based on Reverse Micellar for Bamboo Biomass Hydrolysis in Ionic Liquids

Dan Wang, Zhao Qin

Although ionic liquids (ILs) have emerged as a promising type of solvent for lignocellulose pretreatment, the enzymatic saccharification of regenerated lignocellulose must occur separately due to the toxicity of ILs to cellulase. Therefore, it is important to develop a biocompatible IL-cellulase system in which the IL effectively activates lignocellulosic biomass, and the enzyme maintains high stability. For this purpose, a dual-phase "enzyme-shuttling compartment" was constructed in this study. Polysorbate-80 (also known as Tween 80) was found to form reverse micelles in the isooctane-IL two-liquid phase, acting as a compartment for cellulose and providing a microenvironment that protects and maintains the energetic conformation of the reactive cellulase. The activated bamboo biomass was enzymatically hydrolyzed in 20% (w/v) 1-ethyl-3-methylimidazolium dimethyl phosphate and 50 mM citrate buffer at 50 °C, and a high total reducing sugar (TRS) yield of 71.2% was obtained after 24 h. At the same time, an enzymatic activity of 91.2% was maintained after 24 hours of enzymolysis. The compartment with a diameter of 0.4 to 0.8 μm was observed by TEM. After 16 hours of enzymolysis, the enzymatic activity was maintained up to 93% in the reverse micelles system, compared to 82.3% in the ionic liquid-aqueous solution. Thus, an efficient system with the simultaneous ILs activation and enzymolysis of natural biomass was successfully developed in a one-pot procedure at low temperatures, ensuring large-scale biomass conversion into biofuels and biological products.

502a Developing Models to Understand Performance Trends of Electrochemically-Mediated Carbon Capture Systems

Lauren Clarke, Fikile R. Brushett

Society-wide decarbonization is a grand challenge of the 21st century which will require wide-scale deployment of various carbon-neutral and carbon-negative technologies. Carbon dioxide (CO_2) capture is one promising method as it provides a means to offset emissions from existing fossil-fuel-driven processes and hard-to-decarbonize sectors.[1,2] State-of-the-art approaches for carbon capture typically utilize temperature-driven processes, where CO_2 is captured and released at lower and higher temperatures, respectively.[3,4] While such methods have been demonstrated, even at commercial scales, broad adoption is hampered by drawbacks such as large energetic penalties, high capital costs, complex process designs, and limited scalability. Electrochemically-mediated carbon capture has recently gained interest as this approach can enable higher energetic efficiencies, modular deployment, and operation at ambient conditions. These technologies can also be readily coupled with renewable energy sources, avoiding emissions associated with electricity generation from fossil fuels. However, this is a relatively nascent field of research, and system performance and design principles necessary for feasible operation are not well understood.

At this early stage of development, modeling frameworks hold value in defining achievable performance bounds and describing important relationships between molecular property sets, system design, and operating conditions. Previous analyses have derived thermodynamic relations to determine theoretical system efficiencies, identifying effective material property values which balance tradeoffs and optimize efficiencies.[5–7] These studies have also highlighted how system configuration, a process design characteristic, can significantly increase energy requirements. While this work has led to an improved understanding of performance trends and design criteria, these modeling frameworks do not account for the large energetic penalties which can arise from mass transport, kinetics, ohmics, and other factors. To this end, low-dimensional models can be used as a concise tool to probe the impact of such phenomena, and therefore better characterize electrochemically-mediated carbon capture technologies

In this talk, we describe idealized reactor models which are used to determine cell overpotentials under constant current operation. More specifically, we represent electrodes with continuously stirred-tank reactor and plug-flow reactor approximations, which may serve as lower and upper performance bounds, respectively. We use this modeling framework to explore the impact of molecular/electrolyte properties (e.g., CO_2 solubility), cell design factors (e.g., electrodes, membranes, flowfields), and operating conditions (e.g., current density). One key finding of this work is that overpotentials due to mass transport and ohmics can significantly exceed the minimum energy requirements predicted by thermodynamics. In these scenarios, design criterion derived from thermodynamics need to be expanded to capture operationally relevant properties. We also find that electrode potentials may be asymmetric due to boundary layer phenomena. To further probe this, we apply one-dimensional models to understand the interplay between species transport, heterogeneous and homogeneous reaction rates, and applied current densities within this region. Overall, this work aims to establish connections between material properties, operating conditions, system design factors, and observed performance. Both within the project and more broadly, this modeling approach provides a framework to establish design guidelines for electrochemically-mediated CO_2 separation systems, which can aid in the development and deployment of these technologies.

Acknowledgements

This research was supported by the Alfred P. Sloan Foundation.

References

1. International Energy Agency. *Net Zero by 2050.* https://www.iea.org/reports/net-zero-by-2050 (2021).

2. Allen, M. *et al. IPCC, 2018: Summary for Policymakers.* (2018).

3. Chao, C., Deng, Y., Dewil, R., Baeyens, J. andamp; Fan, X. Post-combustion carbon capture. *Renew. Sustain. Energy Rev.* **138**, 110490 (2021).

4. Fasihi, M., Efimova, O. andamp; Breyer, C. Techno-economic assessment of CO_2 direct air capture plants. *J. Clean. Prod.* **224**, 957–980 (2019).

5. DuBois, D. L., Miedaner, A., Bell, W. andamp; Smart, J. C. Electrochemical Concentration of Carbon Dioxide. in *Electrochemical and electrocatalytic reactions of carbon dioxide* 94–117 (Elsevier Science Publishers B.V., 1993).

6. Shaw, R. A. andamp; Hatton, T. A. Electrochemical CO_2 capture thermodynamics. *Int. J. Greenh. Gas Control* **95**, (2020).

7. Clarke, L. E., Leonard, M. E., Hatton, T. A. andamp; Brushett, F. R. Thermodynamic Modeling of CO2 Separation Systems with Soluble, Redox-Active Capture Species. *Ind. Eng. Chem. Res.* (2022) doi:https://doi.org/10.1021/acs.iecr.1c04185.

502c Microwave Regeneration of CO_2 Solid Sorbent for Energy-Efficient Direct Air Capture

Gyoung Gug Jang, Diana Stamberga, Radu Custelcean, Costas Tsouris

Deployment of energy-efficient regeneration technologies for reagents used in carbon capture is very important in efforts to significantly reduce the high operational cost of direct air capture of CO_2. Guanidine compounds can be used with environmentally friendly aqueous amino acids (e.g., potassium sarcosine) for fast CO_2 capture, and have a strong potential for low-energy CO_2 regeneration. Guanidine compounds [e.g., aqueous 2,6-pyridine-bis(iminoguanidine), PyBIG] react with the CO_2-rich amino acid solution and crystallize as an insoluble carbonate salt [e.g., $PyBIGH_2(CO_3)(H_2O)_4$]. Separation and thermal regeneration of the precipitated guanidine carbonate salt leads to an overall low-energy direct air capture sorbent. Effective CO_2 release from the solid guanidine compound can be achieved by mild heating at 120°C. In this study, to overcome the relatively inefficient traditional conductive heating of crystallized solids, we evaluated the feasibility of microwave heating of CO_2 loaded guanidine ($PyBIGH_2(CO_3)(H_2O)_4$) under various operational conditions such as power, radiation time, and solids mass for efficient CO_2 desorption. We found that microwave heating effectively regenerates guanidine carbonate, which may be facilitated by the water molecules that are co-crystallized with guanidine carbonate. These results indicate that microwave regeneration may be an energy-efficient method for the regeneration of solid sorbents used for direct air capture.

502d Techno-Economic and Life-Cycle Assessment of Electrochemical Captured CO_2 Conversion Process in Monoethanolamine

Suhyun Lee, Woong Choi, Jae Hyung Kim, Yun Jeong Hwang, Jonggeol Na

Severe climate crisis caused by large amounts of CO_2 emissions has led to a global climate agreement about carbon neutrality. Accordingly, efficient carbon capture utilization and storage (CCUS) technology must be developed for the transition to a sustainable low-carbon society. Post-combustion CO_2 capture (PCC) technology based on a 30 wt% aqueous monoethanolamine (MEA) has a mature technology level used on a commercial scale [1]. However, it is an energy-intensive process that consumes more than 57% of total energy for solvent regeneration [2]. In order to overcome this problem, the direct utilization of captured CO_2, which can eliminate the CO_2 separation step, has been proposed as an alternative to the current energy demanding CO_2 separation process. The direct electrochemical conversion via the captured CO_2 reduction reaction (cCO_2RR) proceeds using a CO_2 absorption medium as the electrolyte. Although many studies have been conducted to find the conditions under which the cCO_2RR achieves maximum faradaic efficiency and conversion by changing the type of electrolyte or process operating conditions [3]-[5], there is no precedent that discusses about technological economic prospects of the direct electrochemical conversion process via the cCO_2RR compared to the conventional electrochemical conversion process via the CO_2 supply electrochemical reduction reaction (CO_2RR) process.

Here, we designed two commercial-scale CCU processes, a direct electrochemical conversion process via the cCO_2RR and the conventional electrochemical conversion process via the CO_2RR. And compared them through economic evaluation and environmental analysis. As a result, by comparing the minimum selling price (MSP) of CO product as an economics index, the value of $1.3/kg in the direct electrochemical conversion process and $1.4/kg in the conventional electrochemical conversion process were derived. In other words, our team demonstrated the economic feasibility of the direct electrochemical conversion process with MSP results as low as 7.1% than the conventional electrochemical conversion process. And by performing global sensitivity on current density, overpotential, and faradaic efficiency, it was found that the current density has the greatest influence on the economic feasibility of the entire process. In addition, it was demonstrated that the negative impact on the environment of the direct electrochemical conversion process can be reduced through the life-cycle assessment of the comparative process. The reliable technology economics and environmental analysis reported by our team will be a driving force to take one step further toward the goal of a sustainable society, and this study will be a good guideline used in the future prospect of CCUS technology development.

[1] Khalifa, Omar, et al. 'Modifying absorption process configurations to improve their performance for Post-Combustion CO_2 capture–What have we learned and what is still Missing?' Chemical Engineering Journal 430 (2022): 133096.

[2] Zhao, Bin, et al. 'Enhancing the energetic efficiency of MDEA/PZ-based CO_2 capture technology for a 650 MW power plant: Process improvement.' Applied energy 185 (2017): 362-375.

[3] Lee, Geonhui, et al. 'Electrochemical upgrade of CO_2 from amine capture solution.' Nature Energy 6.1 (2021): 46-53.

[4] Pérez-Gallent, Elena, et al. 'Integrating CO_2 Capture with Electrochemical Conversion Using Amine-Based Capture Solvents as Electrolytes.' Industrial andamp; Engineering Chemistry Research 60.11 (2021): 4269-4278.

[5] Kim, J. H., et al. 'Insensitive cation effect on single-atom Ni catalyst allows selective electrochemical conversion of captured CO_2 in universal media.' (2022).

502e Molecular Dynamics Studies on Separation of CO2/CH4 By the Ionic Liquids Encapsulated ZIF-8

Lei Liu

The utilisation of ZIF-8 membrane with encapsulated ionic liquids (IL@ZIF-8 composite membrane) has been demonstrated as one of the most efficient approaches for CO_2/CH_4 separation. However, compared to pure ZIF-8 membrane, the mechanisms behind the CO_2/CH_4 selectivity enhancement caused by the encapsulation of ILs still remain indistinct. Here we report the progress of understanding the mechanisms from examining the ZIF-8 aperture configuration variation using density functional theory (DFT) and molecular dynamics (MD) simulations. DFT results indicate that the pristine aperture configuration exhibits the best separation performance, and the addition of ILs prevents the apertures from large swing. Subsequently, the effect of IL viscosity on the ZIF-8 configuration variation was investigated, and results show that the pristine aperture configuration is more stabilized by ILs with large viscosity (0-87Cp). Molecular dynamics (MD) simulations show that the selectivity enhancement is because the adsorbed CO_2/CH_4 ratio has been increased when ILs are encapsulated into ZIF-8. Detailed analysis indicates that the diffusion of CO_2/CH_4 molecules follows three different pathways of the hexagonal microstructure of IL@ZIF-8 composite, of which the route to the nearest composite cells (route I) has the highest possibility, and gives the smallest mean diffusion displacement. Lastly, the effect of CO_2/CH_4 ratio in the initial gas mixture has been explored, and we found that CO_2/CH_4 selectivity is related to the concentration of CO_2, where three regions with different slopes have been classified.

454a Upcycling Polyolefin Waste into Performance Elastomers

Weizhong Zou, Alexis Hocken, Bradley Olsen

Recycling represents an important solution to the plastics crisis, it has the potential to substantially reduce waste accumulation while allowing for CO2 reduction by re-use the fossil fuel resources. However, a myriad of challenges, in particular, mixed wastes/multi-plastic parts and dyes and fillers make plastic recycling a problem extremely difficult to solve. Here we developed a new process for polymer upgrading that is based on transforming blends of polyolefins into high performance rubber parts via chemical modification by melt processing. Given our recent studies on the detailed chemistry as well as property predictions for the melt processing of polyolefins, trials to synthesize upcycled rubber from virgin low density, linear low density and high density polyethylene (LDPE, LLDPE, HDPE) are executed as a proper representative of polyethylene (PE) waste. By implementing a melt-modification step with the traditional functionalization and curing, the compatibility among different grades of PE is enhanced while resetting the level of crystallinity via both short chain branching and chain coupling, therefore improves the rubbery behavior at room temperature. With the design of experiments for a thorough understanding on the effects of detailed upcycling formulation along with processing condition on the mechanical properties of upcycled rubbers, our results on LDPE, LLDPE and HDPE pellets show great progress on lowering its original high crystallinity level and the associated yield stress with the proposed upcycling strategy.

454c First-Principles Kinetic Modeling System for the Pyrolysis of Waste Plastics

Pratyush Agarwal, David Tremblay, Zhen Hou, Shu Wang

Plastic waste accumulation in oceans and on land represents a growing concern as we produce about 300 million tonnes of plastic waste every year.[1] Waste plastic is a hydrocarbon rich material that represents a valuable feedstock to recycle into new products, increasing the circularity of plastic use. In this study, a first-principles kinetic modeling system was developed employing Aspen Plus Polymers for the pyrolysis of plastics into chemical feedstocks with the method of moments to track the evolution of inert and radical polymer species in a reactor. Each class of polymer is characterized by representative repeat units. The system automatically generates the free-radical reaction types for the polymers, including random scission, midchain β-scission, cyclic scission, depolymerization, hydrogen abstraction, hydrogen shift, debranching, termination by combination, and termination by disproportionation. A chain-length dependent probability distribution partitions the formation of low-molecular weight oligomer species tracked with one additional set of moments. The moments of the low-molecular weight oligomer species inform the Flory-Schulz distribution to discretize the formation of the specific low molecular weight products from polymer pyrolysis. Additional thermal cracking reactions modeled at the pathway level with power-law kinetics account for secondary conversion of the oligomer products to other observable product moieties. The final model captures both the evolution of the specific low-molecular weight products and the change in bulk polymer properties such as weight-average (M_w) and number-average (M_n) molecular weights, polydispersity index (PDI), and degree of polymerization (DP). Model results compare very well to experimental data reported in literature for common plastics (HDPE, LDPE, PP, PS) as well as their mixtures. This model of waste plastic pyrolysis can be utilized as a starting point for optimization and incorporation of chemical recycling of waste plastics in the circular economy paradigm.

[1] United Nations Environment Program. (n.d.). *Visual feature: Beat plastic pollution.* Retrieved March 20, 2022, from https://www.unep.org/interactives/beat-plastic-pollution/

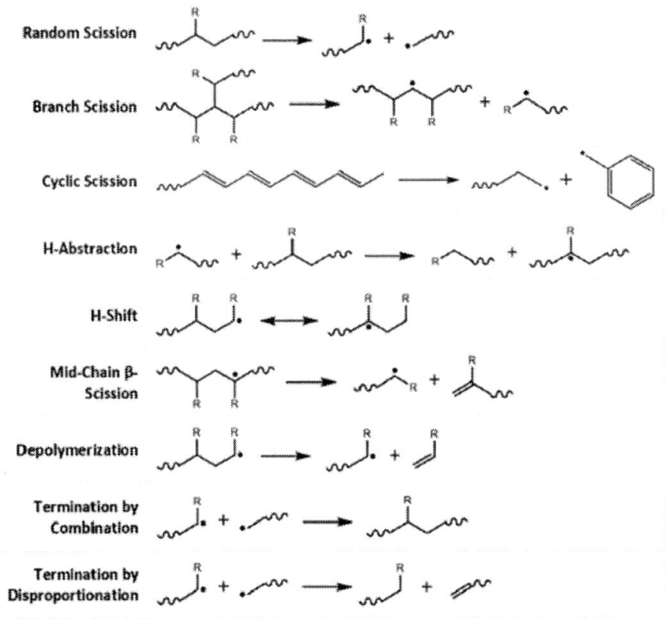

Figure 1. Representative radical reaction mechanisms in polymer pyrolysis

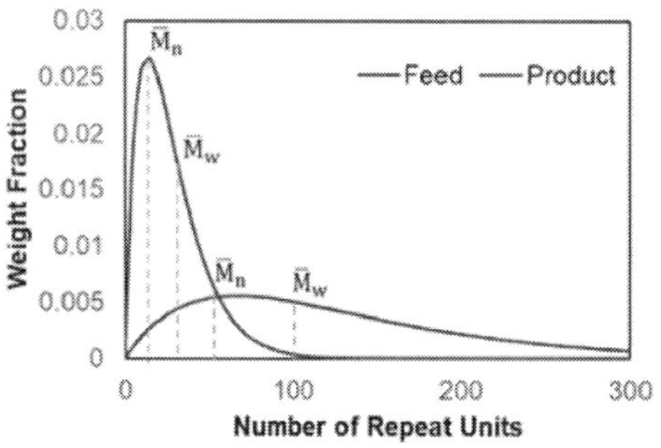

Figure 2. Evolution of polymer properties during polymer pyrolysis

454d Kinetics Insights of Waste Plastic Pyrolysis with Auto-Fragment Modeling (AFM)

Yen-Ting Wang, Matthew Prendergast, William Green

For complex systems such as plastics, it's a huge challenge to build detailed kinetics model to unravel the complicated process due to limited computation resources and time. Therefore, it is necessary to develop new method to tackle the long-lasting issues including exponentially increasing number of isomers and reactions and uncertain accuracy of extrapolation for fitted parameters lumping methods. To overcome difficulties mentioned above, we propose a framework called Auto-Fragment Modeling (AFM) which targets to build detailed kinetic models for heavy hydrocarbons. The method treats large molecules as sliced pieces called fragments and use fragments to construct kinetics model since similar structures will likely to conduct similar reactions. Each piece of fragment represents different part of the original molecules which share common structures among all other species in the system.

This fragment method utilizes Reaction Mechanism Generator (RMG), an open-sourced software, to estimate rate coefficient and thermochemical properties derived from high-accuracy quantum chemistry. Previously we have shown some results for polyethylene (PE) with promising insights by capturing important chemistry and other product information. The framework itself as well as the generated models have been improved with more insights discovered. There are still some technical challenges under development for more accurate predictions on heavy systems.

454e Solvolysis Liquefaction of Low-Rank Coals and Polyolefin Waste-Plastics-Derived Liquid into Mesophase Pitch Intermediates to Manufacture Value-Added Carbon Materials

Wenjia Wang, Karissa Jolley, Mitchell G Nelson, Eric Eddings

As primary packaging materials, polyolefin plastics (PP, HDPE, and LDPE) account for more than 40% of plastic waste discharge but are very hard to reuse. It is more economical to depolymerize the waste polyolefins and convert them into value-added chemicals and fuels. Currently, most attempts at depolymerization of polyolefins involve pyrolysis, which affords waxes and light hydrocarbon mixtures that can be used as an alternative to petroleum-derived liquid fuel. However, the depolymerization of polyolefins into liquid fuel requires an expensive catalytic process with low yield and selectivity, while the polyolefin-derived liquid is economically non-competitive with petroleum gasoline, and the further burning of polyolefin liquid would eventually release the carbon into the atmosphere as CO_2. Therefore, stabilizing the carbons in polyolefin waste plastics in more valuable carbon materials, like carbon fibers, graphite electrodes, or carbon foam, can be a profitable and environmentally-friendly method to utilize the polyolefin waste plastics and reduce plastic waste discharge.

In our previous study, a mild solvolysis liquefaction method converted low-rank bituminous steam coal into a hydrogenated heavy tar intermediate to manufacture value-added mesophase pitch materials with relatively high yield and excellent mesophase formation performance [1]. However, the liquefaction process requires the application of a hydrogen vehicle solvent with a high boiling point, making recycling and reusing the solvent costly. Meanwhile, previous studies also found that the non-catalytic hydrogenolysis of polyolefins leads to an aromatic-rich and naphthenic-rich liquid with high heavy asphaltene and pre-asphaltene fractions. Based on these characteristics of the liquid, it has the potential to be used as a hydrogen vehicle solvent in solvolysis liquefaction. Therefore, depolymerizing polyolefin waste plastics into a hydrogen vehicle solvent, then converting a mixture of polyolefin-derived liquid and low-rank coals into precursors of value-added carbon materials appears very promising.

In this study, various polyolefin waste plastics were hydrogenolyzed into a waste-polyolefin-derived liquid (WPDL) under modified conditions [2]. The obtained WPDL was used as an alternative solvent, in lieu of the pure chemical hydrogen vehicle solvent, in the solvolysis liquefaction of low-rank bituminous steam coal. The liquefaction process was optimized based on the yield and composition of the target product, polyolefin-coal-liquefaction heavy tar (PCL-HT). Solvent fractionation, TGA, FTIR, and ^1H NMR methods were applied to characterize the PCL-HT derived at different conditions. PCL-HT with a high content of heavy fraction and good hydrogen transferability was selected and thermally upgraded into value-added mesophase pitch. The produced mesophase pitch was characterized by DMA, XRD, and polarizing microscopy. The results indicated that the WPDL could replace the pure hydrogen vehicle solvent and refine the composition of the heavy tar derived from the solvolysis liquefaction of low-rank coals. Our research indicates that this method can utilize both polyolefin waste plastics and low-rank coal resources to generate value-added product precursors, especially from different coal-producing areas of the United States.

Reference

[1] Wang W, Preciado I, Malzahn J, Eddings E. Mild Solvolysis Liquefaction of Low-Rank Coal into

a Feedstock of Value-Added Carbon Materials. *2021 AIChE Annual Meeting.* AIChE; 2021.

[2] Wang W, Preciado I, Eddings E. Thermochemical Co-Conversion of Waste Polyolefins with Low-Rank Aromatic-Rich Hydrocarbons into an Intermediate of High-Quality Anisotropic Pitch. *2021 AIChE Annual Meeting.* AIChE; 2021.

66a Mechanism-Based Lignin Bioprocess and Biomaterial Design

Qiang Li, Jorge Arreola Vargas, Cheng Hu, Furong Lin, Bin Yang, Arthur Ragauskas, Joshua Yuan

Lignin is one of the most abundant biopolymers on earth. Despite the abundance, lignin remains primarily a waste stream from paper, pulping, and biorefining industries. The utilization of lignin as a fungible material is fundamental to the sustainable development of our society. Nevertheless, the limited understanding of fundamental structure-property relationship between lignin chemistry and material/product properties have hindered the development lignin processing and materials. Our interdisciplinary team have advanced the understanding how lignin chemistry impacts processibility and material properties. First, we have revealed that lignin processing into smaller molecular weight, monomer-containing, and hydroxyl group-enriched fractions will significantly improve the bioprocessibility. Based on the fundamental understanding, we have designed a 'plug-in' module to transform the conventional pretreatment technologies into integrated biorefinery for both carbohydrate and lignin conversion. In particular, the innovative biorefinery design has enabled lignin processing into PHA to reach market-competitive prices. Second, we have discovered that lignin molecular weight, uniformity, linkage profile, and functional groups could all impact the lignin-derived biomaterials' properties, including carbon fibers, nanoparticles, recyclable plastic blends, and pavement materials. For example, the higher molecular weight, better uniformity, and more linear linkages could improve carbon material properties. The fundamental understanding has guided the lignin process design to derive the type of lignin to achieve highest reported tensile strength for lignin carbon fiber. This new type of lignin can also promote the performance of plastic blends, empowering quality UV-shading biomaterials. The lignin with more condensed structure benefits the stability and size selection for nanoparticles. The lignin functional groups and molecular weight also impact the interaction with asphaltene, in a way that lignin can be tailored to derive asphalt binder modifier with unique properties. Overall, based on the fundamental understanding, we have developed various processes to transform lignin into a precursor for broader biomaterial applications.

66b Reductive Depolymerization of PAA-Fractionated Lignin Via Using Cupmo Catalyst

Anqi Ji, Xianzhi Meng, Jingshun Zhuang, Jiseok Park, Kwang Ho Kim, Gyu Leem, Sung Bong Kang, Arthur Ragauskas, Chang Geun Yoo

Lignin is a key feedstock in biorefinery processes for accomplishing a successful total biomass utilization. It is typically fractionated from carbohydrates such as cellulose and hemicellulose prior to its conversion. However, it is not avoidable to modify its intact structure. Especially, harsh reaction conditions during this fractionation cause unwanted decomposition and condensation, resulting in low end-product yield. Another challenge in lignin valorization processes is the use of expensive catalysts. Various heterogeneous catalysts such as Pt, Pd, and Ru have been applied for effective conversion of lignin; however, these catalysts are costly even though their regeneration is feasible. Therefore, many researchers try to develop earth-abundant metal catalysts to replace current expensive ones.

Cu is one of the earth-abundant metals but is still effective in many thermochemical processes. Successful catalytic hydrodeoxygenation of lignin has been reported in supercritical solvent system such as methanol, using an earth-abundant element Cu-doped porous metal oxide (PMO) as the catalyst, at a relatively mild temperature (300 °C) without addition of external hydrogen. In this single one-pot reaction system, supercritical methanol was catalyzed by the Cu-doped PMO catalyst, releasing H_2 as an in situ liquid syngas precursor. The hydrogen transferred from supercritical solvent to lignin would then result in the hydrogenolysis of phenyl ether bonds coupled with the hydrogenation of aromatic rings.

In this study, peracetic acid (PAA) effectively fractionated lignin preserving relatively high b-O-4 linkage content and minimal condensation due to its mild reaction conditions. CuPMO catalysts were prepared and applied for reductive depolymerization of the fractionated lignins. The fractionated lignins were characterized with two-dimensional heteronuclear single quantum coherence (2D HSQC) nuclear magnetic resonance (NMR), ^{31}P NMR, and gel permeation chromatography (GPC) to understand the impact of PAA on lignin characteristics. The applied catalysts were characterized by inductively coupled plasma (ICP), X-ray photoelectron spectroscopy (XPS), H_2 temperature-programmed reduction (TPR), and NH_3 temperature-programed desorption (TPD). The depolymerized products were analyzed by GPC and gas chromatography mass spectrometry (GC-MS).

66c Identifying Bacterial Lignin Degradation Pathways in *Pseudomonas Putida* KT2440

Zhangyang Xu, Bin Yang

Bacterial lignin depolymerization is believed to be primarily achieved by a secreted enzyme system. Effects of such extracellular enzyme systems on lignin structural changes and depolymerization pathways are still not clearly understood, which remains as a bottleneck in the bacterial lignin bioconversion process. This study investigated lignin depolymerization using an isolated secretome secreted by *Pseudomonas putida* KT2440 that grew on glucose as the only carbon source. Enzyme assays revealed that the secretome harbored oxidase and peroxidase/Mn^{2+}-peroxidase capacity and reached the highest activity at 120 h of the fermentation time. The degradation rate of alkali lignin was found to be only 8.1% by oxidases, but increased to 14.5% with the activation of peroxidase/Mn^{2+}-peroxidase. Gas chromatography-mass spectrometry (GC-MS) and two-dimensional ^{1}H-^{13}C heteronuclear single-quantum coherence (HSQC) NMR analysis revealed that the oxidases exhibited strong C-C bond (β-β, β-5, and β-1) cleavage. The activation of peroxidases enhanced lignin degradation by stimulating C-O bond (β-O-4) cleavage, resulting in increased yields of aromatic monomers and dimers. Further mass spectrometry-based quantitative proteomics measurements suggested that different groups of oxidoreductase enzymes in *P. putida* secretome, including reductases, peroxidases, monooxygenases, dioxygenases, oxidases, and dehydrogenases, contributed to the lignin degradation process. Overall, we discovered that bacterial depolymerization of alkali lignin to vanillin, vanillic acid, and other lignin derived-aromatics involved a series of oxidative cleavage, catalyzed by active DyP-type peroxidase, multicopper oxidase, and other accessory enzymes. These resutls will guide further metabolic engineering design to improve the efficiency of lignin bioconversion.

66d Production of Aromatics from Lignocellulosic Biomass: Vapor-Phase Hydrodeoxygenation of Propylguaiacol over a Bifunctional Molybdenum-Containing Zeolite

Jie Zhu, Matthew Webber, Jamison Watson, Gregg T. Beckham, Yuriy Roman

Lignin, accounting for 15-40 wt% of lignocellulosic biomass, is the most abundant renewable source of aromatic units in nature. Conversion of lignin-derived compounds into aromatic chemicals is a highly desirable target in the biorefinery field.[1] However, selective cleavage of the intrinsic C-C and C-O bonds within lignin remains a long-standing challenge.[2] Early studies have shown that supported Mo and Au catalysts exhibit high selectivity towards C-O bond cleavage in the hydrodeoxygenation (HDO) of lignin-derived monomer.[3, 4] Meanwhile, dealkylation of monomeric lignin, which involves the C-C bond activation that requires higher dissociation energy compared to that of C-O bond, is usually achieved by acidic zeolite.[5,6] Accordingly, it is high desirable to develop efficient catalyst that enables simultaneous cleavage of C-O and C-C bonds for selective production of aromatics from lignocellulosic biomass.

This study investigates the production of aromatics (benzene, toluene and phenol) from lignin-derived propylguaiacol over a bifunctional Mo-containing zeolite. A synergistic effect was observed between the metallic sites and the Brønsted acid sites in promoting efficient C-C and C-O bonds activation to obtain fully deoxygenated aromatics. The close proximity of the metallic sites and the Brønsted acid sites was found to be the key factor for enabling high selectivity towards aromatic and for increasing on-stream stability. In addition, it was revealed a predominant contribution of the acid sites to the cleavage of the alkyl group. Further increase in the acidity of the zeolite support favors the formation of methylated products, resulting in more production of partially deoxygenated aromatics. The reaction network of propylguaiacol HDO was proposed by varying the conversion of propylguaiacol to determine the formation sequence of the primary products. The kinetics for the HDO of propylguaiacol and the intermediates propylphenol, propylbenzene were studied over Mo-containing zeolite. This work will be insightful for the development of bifunctional and bimetallic catalysts for upgrading complex dimeric/oligmeric lignin compounds in the future.

References

[1] C. Zhang and F. Wang, *Acc. Chem. Res.*, **2020**, 53, 470−484.

[2] C. Li, X. Zhao, A. Wang, G. W. Huber and T. Zhang, *Chem. Rev.*, **2015**, 115, 11559−11624.

[3] J. Mao, J. Zhou, Z. Xia, Z. Wang, Z. Xu, W. Xu, P. Yan, K. Liu, X. Guo, Z. Zhang, *ACS Catal.*, **2017**, 7, 695−705.

[4] M. Shetty, E. M. Anderson, W. H. Green and Y. Román-Leshkov, *J. Catal.*, **2019**, 376, 249−257.

[5] D. Verboekend, Y. Liao, W. Schutyser and B. F. Sels, *Green Chem.*, **2016**, 18, 297−306.

[6] Y. Liao, M. d'Halluin, E. Makshina, D. Verboekend and B. F. Sels, *Appl. Catal. B*, **2018**, 234, 117−129.

66e Water-Based Wood Bio-Adhesive Made from Soy Protein Isolate and Depolymerized Lignin

Changle Jiang, Jianli Hu

Soy protein was once a dominant industrial adhesive for bonding wood products. However, it has suffered low bonding strength and poor water resistance. The soy protein adhesive was gradually phased out by the introduction of phenol-formaldehyde and urea-formaldehyde adhesives. Biobased phenolics could be derived from lignin by base-catalyzed depolymerization process under 200 °C. These phenolic compounds, including monomers and oligomers, could be used for crosslinking with soy protein isolate (SPI) molecule, resulting in a protein-lignin adhesive with higher bonding strength as well as better water resistance. After forming crosslinked network with depolymerized lignin, the soy protein-based bio-adhesive has a less water-soluble structure, which helps improve the water resistance. This could be a highly promising way to valorize lignin efficiently and economically. Our results have showed that while adding the untreated lignin decreased the single lap shear strength of the soy protein adhesive, incorporating the depolymerized lignin could enhance the bonding strength. Our bio-adhesive had a single lap shear strength of 912.5 N (equals 1.46 MPa over 625 mm^2 adhesion area), which was at a relatively high strength level when compared with similar soy protein adhesives from the literatures. It is worth noting that 50% of the bio-adhesive solid was from depolymerized lignin, implying the economic benefit due to the use of low-cost biomass derived lignin. More importantly, the solvent was water only. Most of the references had lignin loadings less than 50% in solid content and environment-unfriendly solvents. Additionally, urea was introduced in the adhesive system to unfold the natural tertiary structure of soy protein, which was based on the idea that unfolded protein molecule structure could reduce the hindering effect due to the tertiary structure. The results showed that the urea did not improve the bonding strength.

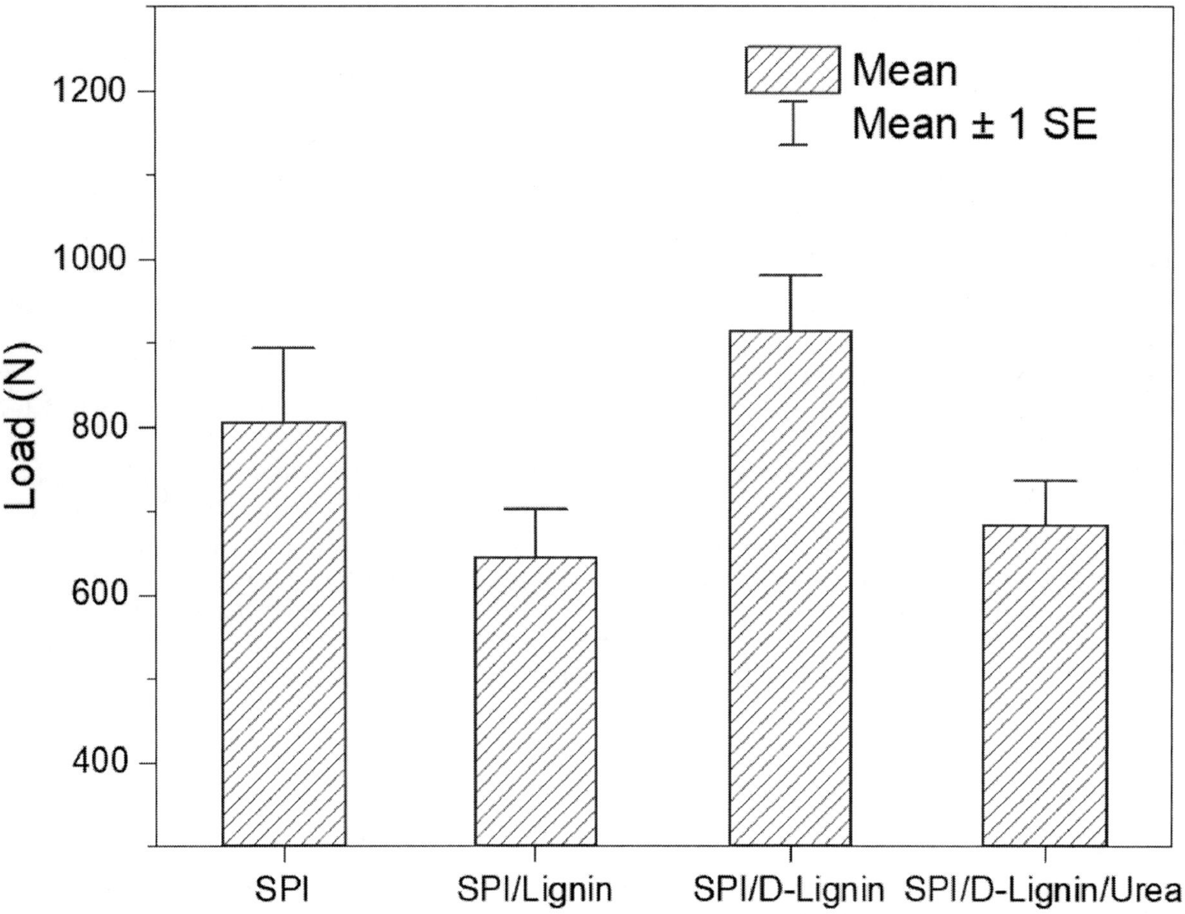

Mean
Mean ± 1 SE
1200
1000
800
600
400
Load (N)
SPI
SPI/Lignin
SPI/D-Lignin
SPI/D-Lignin/Urea

66f Valorization of Lignin: Development Towards a Continuous Fractionation by Precipitation Process

Arulselvan Ponnudurai, Peter Schulze, Andreas Seidel-Morgenstern, Heike Lorenz

The constantly growing global energy demand, concerns about the negative effects of increasing greenhouse gas emissions, and the finite nature of fossil raw materials call for alternative sources, which are low cost, renewable, and environmentally friendly. Lignocellulosic biomass (LCB) is a suitable raw material for carbon-neutral materials and chemical value chains [1]. LCB, can be pretreated by (acetone) organosolv pulping for fractionation into cellulose, hemicellulose and lignin. The latter is an amorphous polymer widely available from various sources [2]. Efficient processes for lignin precipitation from organosolv pulping liquors ("LigniSep") have been developed earlier [3]. However, current technical lignins are hardly industrially utilized due to their complex heterogeneous nature. Tailored narrow lignin fractions could e.g. be used as functional additives or blends in polymer formulations. Possible functionalities of lignin fractions include antioxidant properties, antimicrobial activities and thermal stabilization for materials. Lignin classes of different molecular weight distribution can be precipitated by a stepwise decrease of the acetone concentration of lignin-acetone-water solution [1]. Therefore, the major goal of this work is to design a process for the continuous production of lignin fractions with targeted molecular weight for high-value applications.

Initially stepwise batch fractionation of lignin from different acetone-water compositions was conducted. The obtained fractions showed narrower molecular weight distributions and differ in weight averages and various physicochemical parameters. In this regard, a new High Performance Size Exclusion Chromatography based on acetone as an eluent was developed for determining the molecular weight distribution (MWD) relatively and absolutely (HPSEC-MALS). Through the polarity change in the stepwise fractionation, lignin solubility decreases with reducing the acetone concentration. By enhancing the polarity with increasing water concentration, larger lignin molecules, which indicate non-polar behavior, tend to precipitate. Also in the work of Sameni et al. low molecular weight lignin showed the highest solubility, whereas lignin with the highest aliphatic hydroxyl number had the lowest solubility in an organic solvent [4]. All lignin fractions show antioxidant capacity, whereby a trend towards higher antioxidant activity can be found towards smaller molecular weight lignin fractions. Espinoza–Acosta et al. have reported a similar tendency, where the antioxidant capacity of lignin was higher in fractions with an higher content of phenolic hydroxyl groups [5]. The identification and amount of functional groups was revealed by ^{31}P NMR analysis.

A first concept for an evaporation-based fed-batch process has been derived from the solubility data of fractions obtained by batch processing. Controlling and adjusting the acetone concentration and temperature (by distillation pressure) are essential for the precipitation process. An ATR-FTIR probe is placed into the reactor medium to monitor in-line the acetone concentration and a FBRM probe is used to detect the phase separation of the lignin-rich phase from the solution. The new suggested (semi-)continuous lignin fractionation process, characteristics of the obtained lignin fractions (e.g. MWD-analysis, functional groups, glass temperatures, antioxidant assay etc.) will be presented in the conference contribution along with prospective product applications.

[1] A. Ponnudurai, P. Schulze, A. Seidel-Morgenstern, H. Lorenz (2021): **Separation strategies for valorization of lignin by targeted molecular weight fractionation**, *Poster presented at ISIC 21 – 21st International Symposium on Industrial Crystallization., Virtually*

[2] A. Smit, W. Huijgen (2017): **Effective fractionation of lignocellulose in herbaceous biomass and hardwood using a mild acetone organosolv process**. Green Chem. 19 (22), 5505-5514

[3] P. Schulze, M. Leschinsky, A. Seidel-Morgenstern and H. Lorenz (2019): **Continuous Separation of Lignin from Organosolv Pulping Liquors: Combined Lignin Particle Formation and Solvent Recovery**, *Ind. Eng. Chem. Res.* 58, 9, 3797-3810

[4] J Sameni, S. Krigstin, M.Sain (2017): **Solubility of lignin and acetylated lignin in organic solvent**, *BioRes. 12(1), 1548-1565*

[5] J. Espinoza-Acosta, P. Torrws-Chávez, B. Ramirez-Wong, C. López-Saiz, B. Montano-Leyva (2016): **Antioxidant, antimicrobial and antimutagenic properties of technical lignins and their applications,** *BioRes. 11(2), 5452-5381*

66g Monomers and Biocrude from Hydrothermal Liquefaction of Solvent-Fractionated Lignin

Ronish Shrestha, Feng Cheng, Geoffrey Tompsett, Brent Scheidemantle, Charles M. Cai, Klaus Schmidt-Rohr, Michael T. Timko

Lignin is the second most abundant biological polymer; a natural aromatic biomacromolecule that serves to bind lignocellulose. Together with cellulose and hemicellulose, lignin serves as one of the three major structural biopolymers comprising about 15–40 wt% of lignocellulosic biomass. Its phenolic structure makes it a potential renewable source for organic compounds, especially those containing electron rich aromatic rings. However, lignin's highly recalcitrant nature creates obstacles in its utilization as a renewable alternative source. In contrast to well-studied processes to convert cellulose and hemicellulose, lignin has often been categorized as "waste" with a predicted annual production of~62 M dry tons/year.

Co-solvent enhanced lignocellulosic fractionation (CELF) is a novel biomass pretreatment technique involving dilute acid treatment of biomass in a THF-water mixture and yields a clean lignin byproduct open for valorization. Lignin obtained from this pretreatment process, denoted as CELF lignin, has relatively low molecular weight, few phenolic hydroxyl groups and low content of aryl ether bonds suggesting extensive cleavage of β-O-4 interunit linkages (most dominant) after CELF pretreatment. Hydrothermal liquefaction (HTL) is a promising thermochemical technology for converting lignin and other waste streams in near critical water (T = 240–370 °C and P = 10–30 MPa) into valuable fuel and chemicals. HTL employs hot pressurized water, considered as one of the greenest solvents, favorable for promoting reactions without catalysts. Subcritical water in the form of OH^- and H^+ ions can dissolve and catalyze lignin fragments into phenolic products. From the CELF process, a warm and wet lignin-rich precipitate is generated from a continuous flow system and HTL is ideal for utilizing wet substrates. Hence, we avoid the energy intensive lignin drying step. CELF lignin has thus far never been used as a feed for HTL and the unusual structure of CELF lignin could provide opportunities to further understand the thermal chemistry of lignin depolymerization.

Five different CELF lignin types differentiated by their feed source ranging from agricultural feedstock like pine, corn stover and sugarcane bagasse to hardwoods like poplar and maple wood were selected for this study. Our goal was to characterize, in molecular detail, the structure of CELF lignin and evaluate its thermal reactivity before evaluating HTL and its products for these five CELF lignin. ^{13}C NMR on biomass feeds and their CELF lignin counterparts show significant reduction of aromatic C-O and COO linkers. Almost negligible (<2%) amount of polysaccharides were observed for all CELF lignin samples indicating a clean lignin byproduct obtained via CELF pretreatment. Also, NMR based quantification of lignin monomers provided S/G ratios whose influence was later investigated for downstream HTL degradation. Thermogravimetric analysis (TGA) of CELF lignin samples depict a narrow decomposition temperature range between 300-350 °C. HTL of CELF lignin was performed at 300 °C, 15 MPa for 1 hour with a solid loading of 10 wt%, yielding gaseous, aqueous and char phase products. Biocrude was extracted from the char phase via solvent extraction using acetone. Hardwood derived CELF lignin (poplar and maple) produced the highest biocrude yield of 52 wt%, while bagasse, corn stover and pine CELF lignin had yields in between 40-43 wt%. Major monomers identified in the biocrude using GC-MS included guaiacol, ethylphenol, cresol, ethylguaiacol, syringol, butylated hydroxytoluene, propylguaiacol, and trimethoxybenzene. After quantifying these monomers with FID, Sugarcane bagasse derived CELF lignin produced the highest with 21 mg of monomers per gram of lignin, while pine had the lowest amount with almost no syringol. Thus, HTL of CELF lignin can produce value-added monomers and a biofuel precursor in the form of upgradable biocrude. Understanding the structure and thermal degradation of this novel CELF ligbn nin could also provide insights into feedstock influence and other significant parameters for degradation of all other lignin sources.

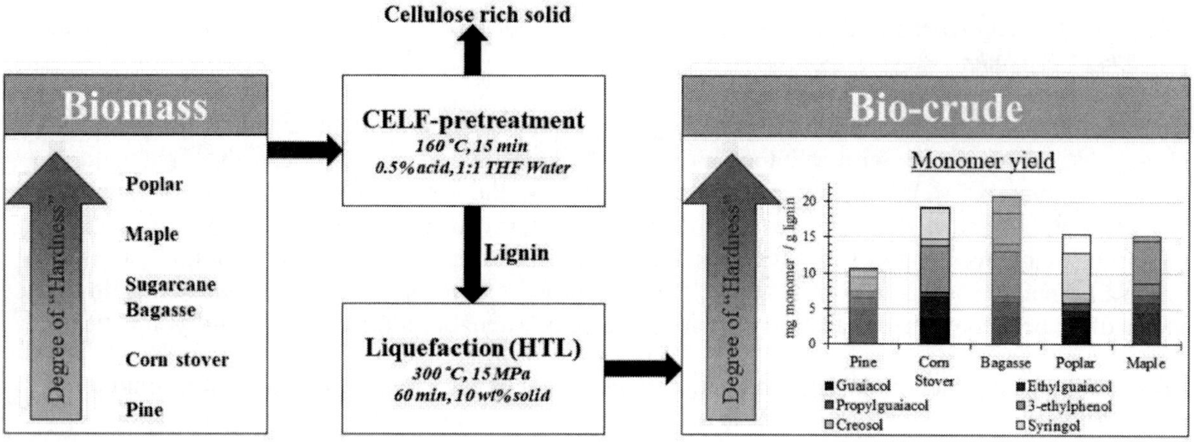
Cellulose rich solid
Biomass
Degree of "Hardness"
Poplar
Maple
Sugarcane Bagasse
Corn stover
Pine
CELF-pretreatment
160 °C, 15 min
0.5% acid, 1:1 THF Water
Lignin
Liquefaction (HTL)
300 °C, 15 MPa
60 min, 10 wt% solid
Bio-crude
Monomer yield
mg monomer / g lignin
Pine
Corn Stover
Bagasse
Poplar
Maple
Guaiacol
Ethylguaiacol
Propylguaiacol
3-ethylphenol
Creosol
Syringol

66h Bioconversion of Rich Containing β-O-4 Linkage Lignin to Medium Chain Length-Polyhydroxyalkanoates

Jorge Arreola Vargas, Bing Xu, Cheng Hu, Xianzhi Meng, Yun-Yan Wang, Arthur Ragauskas, Hugo Oscar Mendez-Acosta, Jose Antonio Perez-Pimienta, Joshua Yuan

Agave tequilana bagasse (ATB) is one of the main residues of the Mexican agro-industry. During the last year by itself more than 800,000 tons of ATB were generated by the Tequila industry according to data from the Tequila Regulatory Council and estimations of the conversion of Agave plant to bagasse (~40%). With tequila demand growing every year, the disposal of this lignocellulosic residue has become an economic and environmental issue for the tequila industry. Although several papers have already been published on the use of the holocellulose fraction of this residue for the production of biofuels and chemicals, scarce information is available on the lignin fraction. Herein, we proposed a sequential pretreatment based on chemical/enzymatic hydrolysis to release the holocellulose and lignin fractions from ATB in different liquid streams. Our work focused on the unexplored lignin fraction, which was fully characterized by gravimetric, 2D ^{1}H–^{13}C HSQC and ^{31}P nuclear magnetic resonance (NMR) and gas chromatography mass spectrometry (GC-MS) analyses. Then, the novel application of this type of lignin for PHA production using engineered *P. putida* strains was initially evaluated in batch and fed-batch fermentation. Thereafter, the fermentation conditions were optimized using a central composite design and optimal conditions were evaluated using three *P. putida* strains. Finally, further insights into the mechanisms of ATB lignin bioconversion were gained by analyzing the chemical composition and structure of lignin before and after fermentation. Results of the chemical structure of the native ATB lignin by NMR analyses revealed that this lignocellulosic residue contains 63% of β-O-4 inter-unit linkages and this percentage only slightly decreased to 50% during our systematic pretreatment, making this lignin an attractive substrate for PHA bioconversion. Exploratory fermentation experiments in batch and fed-batch showed an increase on the PHA titer from 0.09 g/L to 0.39 g/L by using the fed-batch strategy and an engineered *P. putida* strain. Further improvement to 0.76 g/L was possible by using the central composite design to optimize the inoculum, substrate and Nitrogen contents. Finally, the PHA depolymerase gene *phaZ* was knocked out from *P. putida* and evaluated at optimal conditions, achieving a PHA titer of 0.97 g/L. Analysis by NMR and GC-MS confirmed that lignin derivatives like vanillin were totally consumed during fermentation and the β-O-4 inter-unit linkages decreased 30%. Overall, this is the first work reported on the valorization of ATB lignin and sets the precedent for future design of chemical and biological valorization methods for this and other rich containing β-O-4 linkage residues.

66i Fractionation of Depolymerized Lignin Streams Using Tangential Flow Filtration

Dupeng Liu, Asun Oka, Jipeng Yan, Ning Sun

Growing demand for fuels, chemicals, and materials from renewable sources is driving innovation in the conversion of lignocellulosic biomass through biorefineries. Lignin, an environmentally sustainable carbon source, and the second most abundant terrestrial polymer, molecules can vary between 1000 Da and 100,000 Da. With different molecular weight (MW) distributions, they could be used in low-cost carbon fibers, engineering plastics, thermoplastic elastomers, polymer foams, sheet adhesives, and a variety of fuels and chemicals. Precise fractions are regarded as a key procedure to enable efficient and economically viable conversion of lignin to value-added products.[1]

In this talk, the performance of a cascading Tangential Flow Filtration (TFF)[2] for fractionation of the depolymerized lignin streams is presented. TFF systems with different membrane cut-offs (0.5 μm, 25 kDa, 5 kDa, 2 kDa, 1 kDa, to nano) are employed to separate different MW of lignin streams. Different lignin streams made from alkali and ionic liquid pretreatment of corn stover, reductive catalytic fractionation (RCF) of poplar,[3] and oxidative catalytic fractionation (OCF) of poplar,[4] are investigated respectively. A series of analyses such as lignin molecular weight, monomeric sugars, oligosaccharides, and the rheology properties of the streams from different molecular weight cut-off membranes is conducted to understand the lignin fractions, optimize the separation process, and determine the future application. Techno-economic analysis (TEA) results show that a high initial concentration in the liquid stream plays a critical role in the product yield and recovery cost.[5]

1 Ragauskas, Arthur J., et al. 'Lignin valorization: improving lignin processing in the biorefinery.' science 344.6185 (2014): 1246843.

2 Van Reis, Robert, et al. 'High performance tangential flow filtration.' Biotechnology and bioengineering 56.1 (1997): 71-82

3 Anderson, Eric M., et al. 'Flowthrough reductive catalytic fractionation of biomass.' Joule 1.3 (2017): 613-622.

4 Luo, Hao, et al. 'Oxidative catalytic fractionation of lignocellulosic biomass under non-alkaline conditions.' Journal of the American Chemical Society 143.37 (2021): 15462-15470

5 Yan, Jipeng, et al. 'Fractionation of Lignin Streams Using Tangential Flow Filtration.' Industrial andamp; Engineering Chemistry Research (2022).

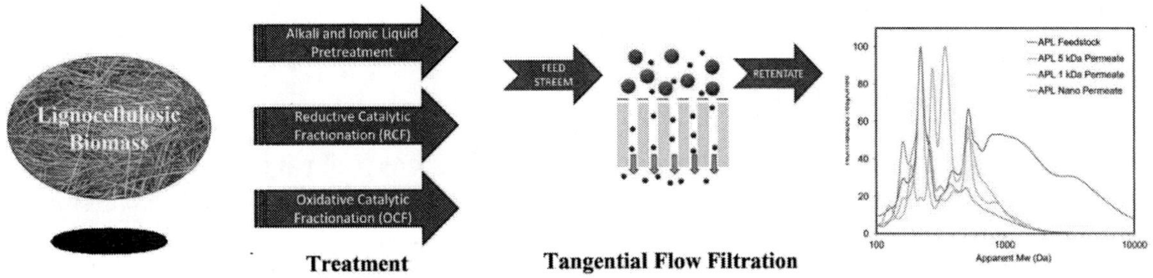

Fig. Tangential flow filtration process was applied to the depolymerized lignin streams fractionation after treatment

66j Reductive Depolymerization of Lignin By Bifunctional Ru-Based Catalysts Supported on Tungstated-Zirconia

Xiaojun Yang, Dequan Xiao, Bin Yang, Maoqi Feng

Catalytic hydrodeoxygenation, which can lead to oxygen-atom removal, ring hydrogenation, and ring-opening, is an effective strategy to depolymerize lignin and increase overall bio-oil conversion yield. In this study, a bifunctional catalyst, low-loading Ru (0.2-0.5 wt.%) supported on tungstated zirconia was evaluated for ring hydrogenation and ring opening of lignin model compounds and technical lignin derived from corn stover. The Ru/W-Zr-O catalyst showed comparable selectivity towards ring-hydrogenation as the Ru/Al_2O_3 catalyst. The selectivity shifted towards aromatics ring-opening when ethanol was used as the solvent. The highest long-chain alkanes bio-oil conversion yielded 90.6% with 79.4% selectivity. By comparing experimental data with the analysis of various calculated bond activation energies of lignin model compounds using density functional theory, we propose that the overall ring-opening reaction mechanism follows the route of "aromatics → cyclic hydrocarbon → linear hydrocarbon". This work provided promise of designing efficient catalysts supported on tungstated zirconia for selective degradation of lignin by fine-tuning the properties of bifunctional catalysts and solvents.

66k Lignin Bioconversion in a Unique Pelletized Cultivation Platform Using *Rhodococcus Opacus* PD630 Simplifies Biomass Harvesting and Enhances Lipid Production

Bing Xu, Qiang Li, Yunqiao Pu, Shangxian Xie, Arthur Ragauskas, Jorge Arreola Vargas, Zhi-Hua Liu, Joshua Yuan

Pelletized liquid cultivation has been widely explored because of its advantages in biomanufacturing, such as easier biomass harvesting, higher product yield, lower medium viscosity and energy consumption. In this study, we discovered that the nonfilamentous bacterium Rhodococcus opacus PD630 could form pellets during the fermentation of alkaline pretreatment liquor (APL) containing lignin as a carbon source. Figure 1 shows pellet formation using APL and no pellet formation with glucose (A=before fermentation, B= after fermentation). This discovery advanced our understanding of bacterium pelletization, as only filamentous fungi and filamentous bacteria were reported to form pellets without the addition of external agents such as flocculants or polymers in previous research. Several factors were investigated to understand how they affect the process of pelletization. Notably, the lipid content in the pellets was much higher than in the scattered bacteria at low nitrogen concentration (<0.5 g/L), under which condition (high carbon to nitrogen ratio) the industrial microbial production for lipids was carried out. Moreover, the highest pellet percentage (~60% of the total biomass) was observed at 30 g/L soluble solid content, an agitation rate of 180 rpm, 1.4 g/L NH_4NO_3, an initial optical density (OD600) of 10, and a centrifugation speed of 6000 rpm. The study opens new avenues to decrease harvesting and cultivation cost as well as energy consumption for microbial fermentation. The physicochemical characteristics of the pellets as well as the chemical changes of the alkaline pretreated liquor after fermentation will be discussed during the talk aiming the elucidation of the pelletization mechanisms.

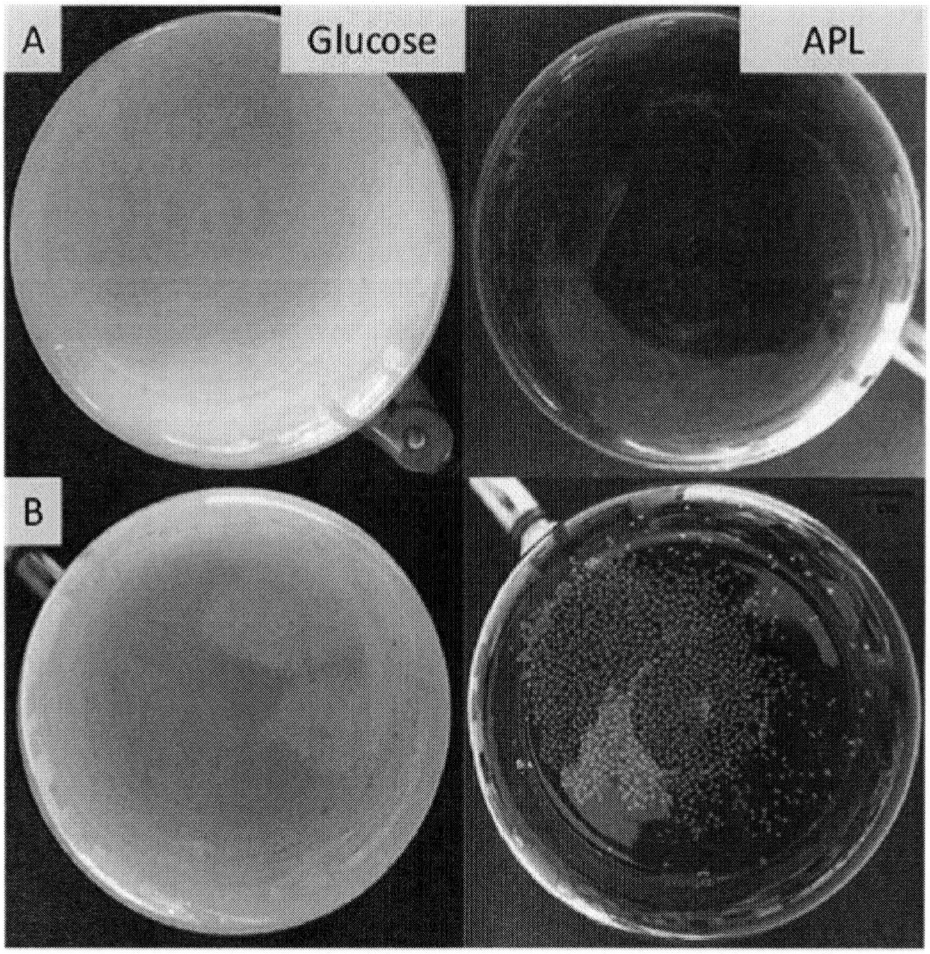
A
Glucose
APL
B

397b Optimization of Nonlinear, Variable-Cost Network Models of the Chemical Manufacturing and Refining Industry

Alkiviadis Skouteris, Ioannis Giannikopoulos, David Allen, Michael Baldea, Mark Stadtherr

In recent years, technological advancements in hydraulic fracturing and horizontal drilling have lead to rapid increases in oil and gas production in the United States, particularly from shale formations [1]. Shale gas is typically rich in natural gas liquids (NGLs), which are comprised of light alkanes other than methane (ethane, propane, butanes and, in smaller amounts, C5+ hydrocarbons), molecules that constitute some of the most important blocks of the chemical manufacturing industry. As a result of the shale gas boom, the production of NGLs has also significantly increased. The increase in the availability of shale gas and associated NGLs has thus provided a unique opportunity to expand the U.S. chemical manufacturing industry [2,3].

The overall aim of this work is to develop a framework for applying systems optimization and design principles to model the industry to provide insight on how to strategically achieve this expansion. This is particularly important given the scale of capital investment involved in industry expansions. Assessing the potential for adoption of new technologies that can make optimal use of NGL resources requires a holistic analysis of the industry. Thus, network models are at the core of such decision making frameworks, due to their ability to capture existing and potential pathways between different manufacturing processes [4].

The petrochemical industry itself is a highly complex, interconnected system of chemical manufacturing and refining processes. Network models of the industry can capture these interconnections and are hence employed in this work instead of a conventional techno-economic comparison of different processes, which does not easily capture industry-wide effects [5]. Network models represent the industry as a directed graph, with nodes corresponding to manufacturing processes, connected via edges that correspond to material flows between the nodes. Nodes are characterized by a set of input-output coefficients – which represent the process material and energy balances – and a net production cost, while edges are characterized by a flow rate. Given the linear nature of the input-output treatment of the transformations taking place at each node, assuming the costs to be constant results in a linear network that can be formulated as a linear program (LP) [6].

In recent work [7], we developed and used an optimization-based network superstructure model of the entire United States, seeking to minimize total industry cost and involving several hundred of the highest-volume chemicals and hundreds of potential processing technologies. A key innovation used was that – in contrast to previous work (e.g. [5,6]) – process costs and material prices were considered to be variable and were allowed to respond to external perturbations in the network structure (e.g., adding a new process). As demonstrated in [7], this can lead to more realistic results but also renders the problem nonlinear. Moreover, it is assumed that the price of a material is driven by the production cost at its largest producer. Since the largest producer may change as a function of the network operating configuration, this leads to a potentially discontinuous dependence of material prices and process costs on the network configuration and process utilizations. To solve the problem, we used a successive linear programming (SLP) approach that alternates between solving a constant-cost LP and a cost-propagation algorithm that updates process costs and material prices based on the LP results.

In recent preliminary work [8], we investigated, and successfully tested on a small network, an alternative approach in which the optimization problem is reformulated as a mixed-integer nonlinear program (MINLP) that can be solved using standard software packages. In this paper, we extend this preliminary work, and explore and compare the SLP and MINLP approaches on a more extensive set of test problems. Furthermore, in order to improve the SLP approach, we describe an alternative cost propagation technique that is capable of more effectively capturing the discontinuous behavior of the price changes due to shifts in the largest producer of a material. This new approach is compared, in terms of both reliability and computational efficiency, to the original approach presented in [7] and to the MINLP approach presented in [8]. These comparisons are based on prototype network examples that capture the network structures encountered in industry, which are then scaled up to more practical cases.

References

[1] EIA, "Natural gas explained: Where our natural gas comes from." https://www.eia.gov/energyexplained/natural-gas/where-our-natural-gas-comes-from.php, 2022. [Online; accessed 07-April-2022].

[2] M. Yang and F. You, "Comparative techno-economic and environmental analysis of ethylene and propylene manufacturing from wet shale gas and naphtha" Industrial andamp; Engineering Chemistry Research, vol. 56, no. 14, pp. 4038-4051, 2017.

[3] J. J. Siirola, 2014. The Impact of Shale Gas in the Chemical Industry. AIChE Journal 60, 810–819.

[4] M. A. Stadtherr and D. F. Rudd, "Systems study of the petrochemical industry," Chemical Engineering Science, vol. 31, no. 11, pp. 1019-1028, 1976.

[5] S. E. DeRosa and D. T. Allen, "Impact of natural gas and natural gas liquids supplies on the United States chemical manufacturing industry: Production cost effects and identification of bottleneck intermediates," ACS Sustainable Chemistry andamp; Engineering, vol. 3, no. 3, pp. 451-459, 2015.

[6] S. E. DeRosa and D. T. Allen, "Impact of new manufacturing technologies on the petrochemical industry in the united states: A methane-to-aromatics case study," Industrial andamp; Engineering Chemistry Research, vol. 55, no. 18, pp. 5366-5372, 2016.

[7] A. Skouteris, I. Giannikopoulos, T. F. Edgar, M. Baldea, D. T. Allen, M. A. Stadtherr, 2021. Systems analysis of natural gas liquid resources for chemical manufacturing: Strategic utilization of ethane. Industrial andamp; Engineering Chemistry Research 60 (33), 12377–12389.

[8] Skouteris, A.; Giannikopoulos, I.; Allen, D. T.; Baldea, M.; Stadtherr, M. A. MINLP framework for systems analysis of the chemical manufacturing industry using network models. Proceedings of the 32nd European Symposium on Computer Aided Process Engineering (ESCAPE32). Toulouse, France, 2022

Acknowledgement

This work is supported in part by the National Science Foundation under Cooperative Agreement No. EEC-1647722 (CISTAR – NSF Engineering Research Center for Innovative and Strategic Transformation of Alkane Resources, http://cistar.us). Any opinions, findings and conclusions or recommendations expressed in this material are those of the author(s) and do not necessarily reflect the views of the National Science Foundation.

397c Techno-Economic Analysis and Life Cycle Assessment of Bio-Based Hydrogen Production from Integrated Dark-Fermentation and Microbial Electrolysis Cells

Xinyu Liu, Adarsh Bafana, Pingping Sun, Amgad Elgowainy, Katherine J. Chou, Alexander Beliaev, Eric Hill, Steven W. Singer, Eric Sundstrom

Renewable Hydrogen (H_2) is a zero-carbon energy carrier that can contribute to the decarbonization of all energy sectors. The U.S. Department of Energy launched the Hydrogen Shot, which seeks to reduce the cost of low-carbon H_2 to $1 per 1 kilogram in 1 decade. Traditionally, H_2 is produced from steam methane reforming (SMR) of fossil natural gas (FNG), emitting 11.5 kg of greenhouse gases (GHG) per kilogram (kg) of H_2 produced on a well-to-wheels basis. To reduce the life cycle GHG emissions associated with the production of low-carbon H_2, renewable biomass resources, such as corn stover, are being considered as a model feedstock for H_2 production.

In this study, we conducted process simulation with Aspen Plus and performed techno-economic analysis (TEA) for bio-based H_2 production using the integrated dark fermentation (DF) - microbial electrolysis cell (MEC) system that converts corn stover to H_2. For dark fermentation, *Clostridium thermocellum* is engineered to decompose both cellulose (C6) and hemicellulose (C5) sugars simultaneously to acetic acid to improve H_2 yield per unit of biomass by weight. In the MEC process, acetic acid in the dark fermentation effluent is converted to H_2 with exoelectrogenic *Geobacter-Shewanella* co-culture. Compared to only 4 moles of H_2/mole sugar (i.e., glucose) from the standalone DF system, the MEC process produces another 8 moles of H_2/mole of starting glucose by utilizing DF by-product (i.e., acetic acid). The integrated DF-MEC system has a theoretical yield of 12 moles of H_2/mole hexose and 6 moles of CO_2. Since the by-product CO_2 is of high purity, we consider captured CO_2 as a potential revenue stream from sales to the beverage industry or as a tax credit through geological storage. Detailed mass and energy balances were obtained from Aspen Plus process model. The reaction parameters employed in the process model were obtained through the concerted efforts of a multi-national laboratory research team, including NREL, LBNL and PNNL, by using lab tested results as well as assumptions made in previous studies.

To perform the TEA analysis of the integrated DF-MEC system, we utilized the Hydrogen Analysis Production Models (H2A) model. Our results indicate that the capital cost of MEC as well as the current density of MEC strongly affect the H_2 production cost. Using recently demonstrated MEC current density of 20 A/m^2, the levelized H_2 production cost is 8.6/kg H_2, assuming 80% DF conversion rate (target) and 90% MEC conversion rate (validated), and assuming additional revenue of $50/metric ton CO_2 due to carbon capture and sequestration (CCS). Increasing the MEC current density to 100 A/m^2 would reduce the cost to $6.3/kg H_2. Further cost reduction opportunities include the continuous enhancement of current density, the use of zero-cost waste feedstock, the reduction of electricity use in MEC, and the reduction of NaOH use in biomass pretreatment.

We utilized the Greenhouse gases, Regulated Emissions, and Energy use in Technologies (GREET®) model to perform the life cycle analysis of bio-H_2 production via biomass DF-MEC. Taking into account the carbon emission reduction from CCS, the life cycle GHG emissions of H_2 production via the integrated system is 5.7 kg/kg H_2, which is higher compared with the 4.8 kg/kg H_2 in the conventional FNG SMR with CCS pathway. This is mainly due to the large use of current fossil-based U.S. electricity generation mix by the integrated DF-MEC system (i.e., 28.3 kWh/kg H_2). If electricity from renewable sources such as wind/solar, is utilized, the integrated DF-MEC system can achieve a net negative life cycle GHG emissions (i.e., -6.8 kg/kg H_2) with CCS. In existing/emerging carbon markets, the reduction in life cycle GHG emissions can result in monetary incentives, which can further reduce the net H_2 production cost.

397e A Simple Framework to Explore the Financial and Technical Challenges of Decarbonization and Sustainability

Carlos Villa, Narayan Ramesh

Decarbonization can be accomplished through the use of green energy, by sequestering CO_2, or by simply developing process technologies that avoid generating CO_2 in the first place. The conversion to alternate technologies requires capital to build the new plants and replace the existing asset base. Policy makers around the globe have proposed a tax on CO_2 as a means to fund the capital and accelerate decarbonization. In this presentation, we make use of several limiting cases to study the interplay between a CO_2 tax, the capital required to decarbonize, the return on capital, the operating expenses and an increased productivity based on operating an asset for many years. Sensitivities of various parameters for the limiting cases and the value chain implications are explored. The problem of decarbonization is complicated due to the complexity, the large capital required and the critical role policy plays in the decarbonization journey. These limiting cases provide a means to explore sensitivities and intuitively design approaches that can help us achieve success.

397f Evaluation and Optimization of Decarbonization Strategies in Indian Steel Industry for Sustainable Scenarios

Sydney Johnson, Lingyan Deng, Emre Gençer

India is of particular interest in the context of climate change initiatives due to the immense economic and population growth expected between 2020 to 2040[1]. The urban population is expected to increase by 270 million people and building expansion could result in residential floor space nearly tripling. Out of the industrial sectors, iron and steel is the most emissions-intensive in the country and accounts for 12% of national emissions. Steel production is expected to increase by at least 100 mega-tonnes, driven by consumption in construction, infrastructure, and capital goods[2,3]. Unlike other countries, India relies heavily on coal for iron and steel-making; this fossil fuel represents 88% of energy consumption. According to the Ministry of Steel, without implementing decarbonization strategies in the sector, 2030 CO_2 emissions could be almost 200 Mt greater than without intervention[4]. Modeling these decarbonization strategies in terms of cost and emissions on a plant and systems level is critical to understanding the impacts and necessary changes for climate mitigation targets.

The cost and emission analysis are developed and evaluated using the Sustainable Energy Systems Analysis Modelling Environment (SESAME). SESAME is a software tool that conducts life cycle, techno-economic, and systems-level analysis for power, transportation, and industrial sectors[5,6]. Specifically, the SESAME Industry Model calculates gate-to-gate CO_2 emissions for the raw material preparation, ironmaking, and steelmaking steps. A combination of literature and industry sources provide material and energy inputs for the plant-level carbon balance. SESAME currently includes the following steel production routes: BF-BOF (and with CCS), HIsarna-BOF (and with CCS), COREX-BOF (and with CCS), DRI-EAF (Coal, Natural Gas, Electrolytic Hydrogen), and scrap-based EAF. BF-BOF and DRI-EAF (Coal) are the two dominant pathways for producing steel in India. Costs are based on capital and operational expenditures sourced from literature and industry reports and scaled accordingly. SESAME also analyzes the direct emissions and costs for the Indian iron and steel industry. The systems level analysis assists in evaluating and optimizing production strategies for steel demand based on EPPA and TERI projections[2,7]. The optimization model is mixed-integer linear program that presents a portfolio of strategies based on minimized cost. Overall, the SESAME Industry Model is a user-facing tool that assists in estimating emissions andamp; costs for decarbonization for the iron and steel industry within India and beyond.

The current, preliminary emission analysis indicates a combination of decarbonization options exists for India's steel industry. For BF-BOF, smelting alternatives offer at least a 10% reduction by switching to the more energy-efficient technologies. Smelting reduction technologies also benefit from being combined with other decarbonization methods like carbon capture. With a post-combustion system of MEA solvent and natural gas for regeneration fuel, emissions are decreased by at least 50%. Within primary steelmaking, DRI-EAF alternatives, the emission results yield an almost 60% reduction for natural gas and a 1% increase in emissions for electrolytic hydrogen when relying on the main grid[8]. By introducing more renewable sources of energy, H_2-DRI reduces emissions by more than 80%. Further analysis will build upon the current emission results by including cost and system-level evaluation of the impact of decarbonization strategies in the Indian iron and steel industry.

Sources

1. "India Energy Outlook 2021 – Analysis." Accessed: Feb. 10, 2021. [Online]. Available: https://www.iea.org/reports/india-energy-outlook-2021

2. Hall, T. Spencer, and S. Kumar, "Towards-a-Low-Carbon-Steel-Sector-Report.pdf," 2020. https://shaktifoundation.in/wp-content/uploads/2020/01/Towards-a-Low-Carbon-Steel-Sector-Report.pdf.
3. Gupta, C. Apran, and Federation of Indian Chambers of, "Indian Secondary Steel Industry Opportunities and Challenges." https://ficci.in/publication.asp?spid=20782
4. Ministry of Steel, "Template on INDC-Mitigation," 2015 2014. https://steel.gov.in/sites/default/files/TEMPLATES-%20MITIGATION_0.pdf
5. Gençer, E.; Torkamani, S.; Miller, I.; Wu, T. W.; O'Sullivan, F. Sustainable Energy System Analysis Modeling Environment: Analyzing Life Cycle Emissions of the Energy Transition. Applied Energy 2020, 277, 115550. https://doi.org/10.1016/j.apenergy.2020.115550.

6. Gençer, E. et al. SESAME https://sesame.mit.edu/.

7. Paltsev, S.; Gurgel, A.; Morris, J.; Chen, H.; Dey, S.; Marwah, S. Economic Analysis of the Hard-to-Abate Sectors in India. 34.

8. IEA (2020), *Tracking Power 2020*, IEA, Paris https://www.iea.org/reports/tracking-power-2020

397g Techno-Economic Feasibility Analysis of Liquid Metal-Based Methane Pyrolysis Process

Semie Kim, Young-Il Lim, Uen-Do Lee, Youn-Bae Kang, Sungwon Kim

Efforts are being made to move forward from fossil fuels to renewable energy, coping with climate change. Hydrogen produced from renewable energy can be used without generating greenhouse gases. Hydrogen is produced by steam methane reforming (SMR), oil/naphtha reforming, coal gasification, water electrolysis, and methane pyrolysis. Non-oxidative methane pyrolysis is a CO_2-free H_2 production technology that produces solid carbon instead of CO_2. The reaction of methane decomposition is an endothermic high-temperature process that requires heat. The heat is generally supplied using coal, natural gas, and solar heat. Molten-metal (MM) based CH_4 pyrolysis has advantages over solid catalytic reactors because catalyst deactivation by coke is prevented. The downstream processing of MM-based CH_4 pyrolysis is simpler than SMR because H_2 and solid carbon are produced.

In this study, process modeling for MM-based methane pyrolysis with different heating sources and carbon separation were presented to evaluate the energy, CO_2 emissions, and economics. Economic indicators such as total capital investment (TCI), total production cost (TPC), return on investment (ROI), and payback period (PBP) were also assessed.

397h Process Design and Economic Analysis for the Retrofitting of a Hydrogen Plant with CO_2 Capture

Joohwa Lee, Haryn Park, Seokwon Yun, Jin-Kuk KIM

Process Design and Economic Analysis for the Retrofitting of a Hydrogen Plant with CO_2 capture

Joohwa Lee[1], Haryn Park[1], Seokwon Yun[2], Jin-Kuk Kim*[1]

[1]Hanyang University, Korea, Republic of

[2]SK Innovation, Korea, Republic of

*Email: jinkukkim@hanyang.ac.kr

It is much more economic to use fossil fuels than to use renewable energy for the commercial-scale production of hydrogen. However, simply using fossil fuel is not a clean method due to the emission of CO_2, so the introduction of CO_2 capture process would be the most realistic way to reduce CO_2 emissions for sustainable production of hydrogen. Among various CO_2 capture methods, absorption-based CO_2 capture process is widely used for a large-scale plant. However, integrating the capture process into the hydrogen plant requires additional capital investment and operating expenditure, which leads to a significant increase in the production cost of hydrogen. Although many studies have been conducted to introduce the CO_2 capture process to industrial plants, systematic and comprehensive economic evaluation has not been made for the retrofitting of a hydrogen plant with a capture plant.

In this study, a baseline hydrogen plant and CO_2 capture process model is created using a commercial simulator Aspen HYSYS®, while configurational changes associated with CO_2 capture options are also examined and different kinds of retrofit strategies were considered for all possible CO_2 capture options in a holistic manner. The accuracy and applicability of the simulation model developed is validated with literature data. From the simulation results, an economic evaluation study is carried out to estimate the Cost of Hydrogen, CO_2 Capture and Avoidance Cost with breakdown and sensitivity analysis presented. Therefore, the cost-effective strategy for the retrofitting CO_2 capture process into a hydrogen plant is identified, and the optimal configuration and its operating conditions can be systematically determined. Case studies provide conceptual insights and practical design strategies for the retrofitting CO_2 capture into the hydrogen plant.

Acknowledgement

This work was supported by the National Research Foundation of Korea (NRF) grant funded by the Korea government (MSIT) (No. 2019R1A2C2002263) and by the Korea Institute of Energy Technology Evaluation and Planning (KETEP) and the Ministry of Trade, Industry andamp; Energy (MOTIE) of the Republic of Korea (No. 20214710100060).

196a Multiscale Inverse Design of High-Performance Ionic Liquid Solvents for High-GWP Refrigerant Separation

Mohammed Sadaf Monjur, Ashfaq Iftakher, Faruque Hasan

Hydrofluorocarbons (HFCs) are predominantly used as refrigerants, but have high greenhouse gas potential (GWP). Currently, there are a million kilograms of HFCs that are being used worldwide, which need to be recycled and reused. Recycling these HFCs is important to prevent the release of unwanted high-GWP gases into the environment. However, due to the azeotropic/near azeotropic nature of some of the HFC mixtures, sustainable separation of HFCs using conventional distillation is nearly impossible. To that end, ionic liquids (IL) have shown great potential as a solvent in extractive distillation (ED)-based separation of HFCs [1,2]. The choice of IL plays an important role in ED-based process performance and the attainable separation quality [3]. Most ILs are liquid at room temperature which can be synthesized by combining anions and cations. Theoretically, there are millions of hypothetical ILs that can be constructed, which gives them the name "designer solvents". To expedite the search for the optimal IL for separating refrigerant blends, it is imperative to construct a multiscale process synthesis, optimization, and material screening framework.

In this work, we propose a novel "inverse design" approach for materials discovery and illustrate the efficacy of this approach in finding the most desired ILs for ED-based separation of HFCs. Our proposed approach is able to decouple the multiscale simultaneous materials-process screening problem into two separate subproblems using feasibility maps [4]. These feasibility maps are obtained by solving the full-scale process synthesis problems for numerous hypothetical ILs, thereby allowing parallel computing and fast identification of the property domains of interest [5]. We extract the optimal domain of thermodynamic and physical properties of ILs based on our recent work where we employ SPICE_ED for ED-based intensified separation schemes of refrigerant blends [3,6]. Next, we obtain a reliable quantitative structure-property relationship (QSPR) of ILs by developing an artificial neural network (ANN). To overcome limited data availability for ANN training, we use COSMO-RS-based simulations to predict the thermodynamic and physical properties that include Henry's constant, density, and specific heat capacity, among others. After establishing confidence in the derived QSPR by performing cross-validation, we transform the trained ANN model into an discrete optimization problem whose solution provides new IL structures with desired properties that match with the feasibility maps. We continue generating new structures until the ANN-predicted properties match with COSMO-RS predicted properties within sufficient accuracy. We apply the design framework to identify new ILs for separating R-410A (50 wt% mixture of R-32 and R-125). We anticipate that such a dynamically updated closed-loop inverse design procedure could be an exciting pathway for identifying novel molecular structures for separation applications beyond extractive distillations.

Keywords: Inverse design, Machine Learning, Hydrofluorocarbons, Ionic Liquids, Extractive Distillation, Process Synthesis.

References:

[1] Shiflett, Mark B.; Yokozeki, A. Solubility and diffusivity of hydrofluorocarbons in room-temperature ionic liquids. *AIChE Journal (2006), 52 (3),* 1205-1219.

[2] Baca, K.R., Olsen, G.M., Matamoros Valenciano, L., Bennett, M.G., Haggard, D.M., Befort, B.J., Garciadiego, A., Dowling, A.W., Maginn, E.J. and Shiflett, M.B., 2021. Phase Equilibria and Diffusivities of HFC-32 and HFC-125 in Ionic Liquids for the Separation of R-410A. *ACS*

Sustainable Chemistry andamp; Engineering, 10(2), pp.816-830.

[3] Monjur, M.S., Iftakher, A. and Hasan, M. M. F., 2022. Separation Process Synthesis for High-GWP Refrigerant Mixtures: Extractive Distillation using Ionic Liquids. *Industrial andamp; Engineering Chemistry Research, 61(12),* pp.4390–4406.

[4] Iyer, S. S.; Hasan, M. M. F. Mapping the Material Property Space for Feasible Process Operation: Application to Combined Natural Gas Separation and Storage. *Industrial andamp; Engineering Chemistry Research, 2019, 58*(24), pp.10455-10465.

[5] Monjur, M.S., Iftakher, A. and Hasan, M. M. F., Ionic Liquid-based Energy Efficient Separation Pathways for High-GWP Refrigerant Mixtures: Solvent Screening and Process Intensification. *Under Review.*

[6] Monjur, M.S., Iftakher, A. and Hasan, M. M. F., Sustainable Process Intensification of Refrigerant Mixture Separation and Management: A Multiscale Material Screening and Process Design Approach. *Proceedings of the 14th International Symposium on Process Systems Engineering – PSE 2021+, Accepted.*

196b Measuring Henry's Law Constants and Infinite Dilution Activity Coefficients of Gases and Liquids in Ionic Liquids Using Gas Chromatography for Olefin/Paraffin Separations

Kimberly Bourland, Joan Brennecke

Ionic liquids (ILs) are widely studied for gas separations because of their low vapor pressures and high gas solubilities. Volumetric and gravimetric apparatuses are used to accurately determine solubilities of gases in ILs; however, they can be incredibly time consuming. A faster method that can be used to screen ionic liquid for specific separation applications is gas chromatography (GC). It yields Henry's Law Constants ($\mathscr{H}$, for gases) and infinite dilution activity coefficients (γ^{∞}, for liquids), which are good measures of the IL-solute interactions. As a result, the GC method can be used to quickly and easily screen numerous ILs for the strength of their interactions with multiple gases and liquids. We present $\mathscr{H}$ and γ^{∞} values of various solutes, including ethane, ethylene, propane, propylene, a family of C_5-C_8 linear alkanes and alkenes, and aromatics in imidazolium ionic liquids, including 1-ethyl-3-methylimidazolium bis(trifluoromethylsulfonyl)imide ([emim][Tf_2N]), and 1-allyl-3-methylimidazolium bis(trifluoromethylsulfonyl)imide ([allylmim][Tf_2N]). Finally, we show the dramatic changes in $\mathscr{H}$ and γ^{∞} values one can achieve for olefins when mixing these ionic liquids with silver salts (Ag bis(trifluoromethylsulfonyl)imide) at concentrations up to 2 M, in order to improve the selectivity when using ILs for of olefin/paraffin separations.

196c New Dimensions in Fabricating Ionic Polymers of Intrinsic Microporosity for Advanced Membrane Gas Separation

Shuangjiang Luo, Wei Xie, Zhili Cai, Can Wang, Peijun Zheng

Next generations of high-performance membranes featuring fast and selective transport combined with robust chemical, mechanical, and thermal stability are highly desired for various energy-demanding industrial gas separation processes. Gas transport properties (i.e., permeability and selectivity) of synthetic polymer membranes suffer a ubiquitous trade-off, known as the Robeson's upper bounds, which originates mainly from the difficulty in simultaneously increasing fractional free volume while manipulating the microcavity size distribution. On this basis, novel polymer architecture with high fractional free volume (FFV) and well-defined micropore size distribution is highly demanded for the fabrication of high-performance gas separation membranes.

Ionic liquids (ILs) are molten salts at ambient temperature and have attracted substantial attention in membrane-mediated gas separation due to their characteristic electrostatic interaction, favorable gas affinity, and abundant structure variations. This talk focuses on the macromolecular design, synthesis, characterization, and separation performance of a novel family of ionic polymers of intrinsic microporosity (IPIM) as gas separation membranes, which have shown promising separation performance for various gas pairs [1,2]. The functionalization of polymers of intrinsic microporosity with ionic species provides additional tunability towards the microcavity architecture through the interchain electrostatic interactions and the counterions. The wealthy structural hierarchy and the versatile chemical compositions of the ionic functionalization offer unique opportunities for generating tailored "molecular cavities" in the IPIM membranes, in which hierarchically architecture is developed with microcavities favoring fast gas transport and ultra-micropores boosting gas sieving. These novel IPIM membranes will be discussed in terms of their synthesis, characterization, and gas separation properties, focusing on how to finely tune the hierarchical microcavities via introducing various ionic substituents, manipulating the degree of ionization, and varying molecular configurations in the membranes. The fundamental correlations between the microscopic structures of these new ionic polymer of intrinsic microporosity and macroscopic membrane transport properties will be presented.

References

1. Xie, Y. Jiao, Z. Cai, H. Liu, L. Gong, W. Lai, L. Shan, S. Luo*. Highly Selective Benzimidazole-Based Polyimide/Ionic Polyimide Membranes for Pure- and Mixed-Gas CO_2/CH_4 Separation. *Separation and Purification Technology*, **2022**, *282*, 120091.

2. Z. Cai, Y. Liu, C. Wang, W. Xie, Y. Jiao, L. Shan, P. Gao, S. Luo*. Ladder Polymers of Intrinsic Microporosity from Superacid-Catalyzed Friedel-Crafts Polymerization for Membrane Gas Separation, *Journal of Membrane Science*, **2022**, *644*, 120155.

196d Ionic Liquids Based Membranes for Gas Separation

Lu Bai

A large number of undesired gases (e.g. CO_2, NH_3) inevitably emitted from the industries, causing serious environmental problems, like greenhouse effect, fog and haze. Meanwhile, CO_2 and NH_3 are basic raw materials to produce important chemicals, such as dimethyl carbonate, ethylene glycol and urea. Therefore, developing novel technologies for efficient gas separation is crucial. Among available methods, membrane gas separation is promising alternative due to high energy efficiency, small footprint and easy operation. However, the trade-off between permeability and selectivity of the membrane materials is still challenged. Ionic liquids (ILs) as a novel medium possess negligible vapor pressure, designable property and high gas affinity, making them widely applied in CO_2 capture and NH_3 separation. The unique properties of ILs incorporated into membranes will significantly influence membrane microstructures and separation performances, thus breaking through the trade-off limitation.

In this work, strategies including construction of nanochannels, regulation of interfacial properties and enhancement of gas affinity are used to design and fabricate novel functional ILs based membranes. The IL/polymer blended membranes were developed for efficient CO_2 capture. Due to the concurrent effect of the soft chain of Pebax copolymer and the inherent fluid nature and high CO_2 absorption ability of ILs, the CO_2 separation performance of IL/polymer blended membranes were improved. It was found that the continuous fluid phase was formed in the blended membranes with high IL content, enhancing the CO_2 solubility and diffusivity. Besides, ILs with ammonia interaction sites were introduced into sulfonated copolymer to fabricate hybrid membranes for NH_3 separation. The NH_3 permeability and selectivity of the hybrid membrane was simultaneously improved. The NH_3 permeability is up to 3248 Barrer and the NH_3/N_2 ideal selectivity reaches 1662. The various characterizations and simulations revealed that continuous ionic domains with well-distributed ILs were formed in the ILs/Nexar hybrid membranes, contributing to forming interconnected channels for enhanced NH_3 transport. In this case, addition of the ILs into the Nexar matrix significantly improves NH_3 separation performance.

196e Efficient Fixation of CO_2 into Cyclic Carbonate Catalyzed By Ionic Liquids in Microchannels

Feng Huo, Yu Chen, Jiayuan Yu, Yiqian Yang, Chunshan Li

Carbon dioxide (CO_2) is a greenhouse gas, the increase of CO_2 concentration will not only lead to global warming, but will also cause frequent extreme weather events.[1] However, CO_2 is still an abundant C1 resource, and converting CO_2 into chemicals with high added value is an attractive approach to carbon-neutral. The conversion of CO_2 with epoxide to cyclic carbonate, which can be used as lithium ion battery components, is an atom-economic and effective CO_2 utilization route.[2] Ionic liquids (ILs) have gained widespread attention as green solvents and new catalytic systems due to their excellent chemical and thermal stability.[3] However, the cycloaddition of CO_2 with epoxides is a typical gas-liquid multiphase reaction involving both gas-liquid mass transfer and chemical reaction, which is commonly carried out in batch reactors for a long time under high temperature and pressure conditions with low efficiency.[4] In recent years, micro-reactor technology has been developed as a novel approach for process intensification, and could hopefully enhance this CO_2 fixation reaction. The interfacial area (a) and the mass transfer coefficient of liquid-phase (k_L) are enhanced in microreactors due to the small channel size.[5]

Cyclic carbonate was synthesized by CO_2 cycloaddition with ionic liquid as catalyst in a 2mm microchannel reactor. The impact of mass transfer and intrinsic kinetics in the microchannel reactor on the ionic liquid-catalyzed coupling reaction of carbon dioxide (CO_2) and propylene oxide (PO) for the synthesis of propylene carbonate (PC) were investigated by examining the influence of various catalysts, catalyst concentration, reaction temperature, CO_2/PO molar ratio, operating pressure, and residence time on the system performance.

A series of imidazole based ionic liquids catalysts were screened for the synthesis of PC, and the discoveries revealed that 1-hexyl-3-methylimidazolium bromide (HMIMBr) with longer alkyl chains and stronger nucleophilic ability had the best catalytic activity. At 125°C and 2.5 MPa, PC yields of more than 99% can be achieved in a 2mm micro-reactor with a residence time of 5 minutes. Subsequently, the kinetics of CO_2 fixation into cyclic carbonate catalyzed by [Hmim]Br were investigated in the microchannel reactor. The activation energy for [Hmim]Br at the temperature range from 65°C to 125°C was calculated as 56.86KJ/mol and 62.55 kJ/mol for 1mm and 2mm micro-reactor, respectively, and the reaction rate showed a first-order dependence on PO. The decrease in reactor size from 2 mm to 1 mm inner diameter resulted in an approximately double improvement in catalytic performance and a slight decrease in reaction activation energy, which was mainly attributed to the increase in mass transfer efficiency. And the specific area of microchannel was found to range from 1000 to 10000 m^2/m^3, and the volumetric mass transfer coefficient (k_La) was 1 to 2 orders of magnitude higher than that of ordinary gas-liquid contactors. The present work was compared with the results reported in the literature and is summarized in Fig.1(c). In general, the reactions of CO_2 and epoxide are usually carried out in batch reactors, either under stringent operating conditions or with poor results. Thus, compared to conventional reactors, micro-reactors weaken the mass transfer limitations, but for the typical fast inversion process of CO_2 transition to cyclic carbonate, the reaction performance is still mainly controlled by mass transfer. Increases in operating temperature, pressure and catalyst concentration optimize the reaction, and residence time can be drastically reduced from several hours in a conventional batch reactor to about a few minutes in a microreactor.

In brief, despite the mass transfer limitations in microchannel reactors can be significantly attenuated compared to conventional reactors, the reaction performance of CO_2 and PO is still considerably affected by mass transfer. This provides new opportunities for enhanced gas-liquid reactions with typical mass transfer control and for CO_2 utilization to achieve carbon-neutral, but the development of more efficient microreactors remains challenging. For highly efficient microreactor development, more dedicated research is needed to reach a compromise between mass transfer and intrinsic kinetics.

Fig1. (a) Schematic of the experimental setup for the cycloaddition of PO and CO_2. (b) Effect of the catalyst on the yield and STY. (c) Diagram of the Arrhenius curve of ln k_{obs} versus 1/T for 1 mm and 2 mm microchannel reactors. Reaction conditions: 125°C, Catalyst/H_2O/PO(mol)=0.14/0.25/1, 2.5 MPa, Q_G=176ml/min(STP), Q_L=0.5ml/min. (d) Comparison of microchannel reactor with other traditional reactors.

This work was supported by the National Key Projects for Fundamental Research and Development of China (2020YFA0710202).

Reference:

1. Cui G, Wang J, Zhang S. Active chemisorption sites in functionalized ionic liquids for carbon capture. *Chemical Society Reviews*. 2016;45(15):4307-4339.

2. He M, Sun Y, Han B. Green carbon science: scientific basis for integrating carbon resource processing, utilization, and recycling. *Angew Chem Int Ed Engl*. 2013;52(37):9620-9633.

3. Zhang S, Sun J, Zhang X, Xin J, Miao Q, Wang J. Ionic liquid-based green processes for energy production. *Chemical Society Reviews*. 2015;46(3):7838-7869.

4. Chaemchuen S, Xiao X, Ghadamyari M, et al. Robust and efficient catalyst derived from bimetallic Zn/Co zeolitic imidazolate frameworks for CO_2 *Journal of Catalysis*. 2019;370:38-45.

5. Ma D, Zhu C, Fu T, Ma Y, Yuan X. Performance and pressure drop of CO_2 absorption into task-specific and halide-free ionic liquids in a microchannel. *AIChE J*. 2022;e17613. doi:10.1002/aic.17613.

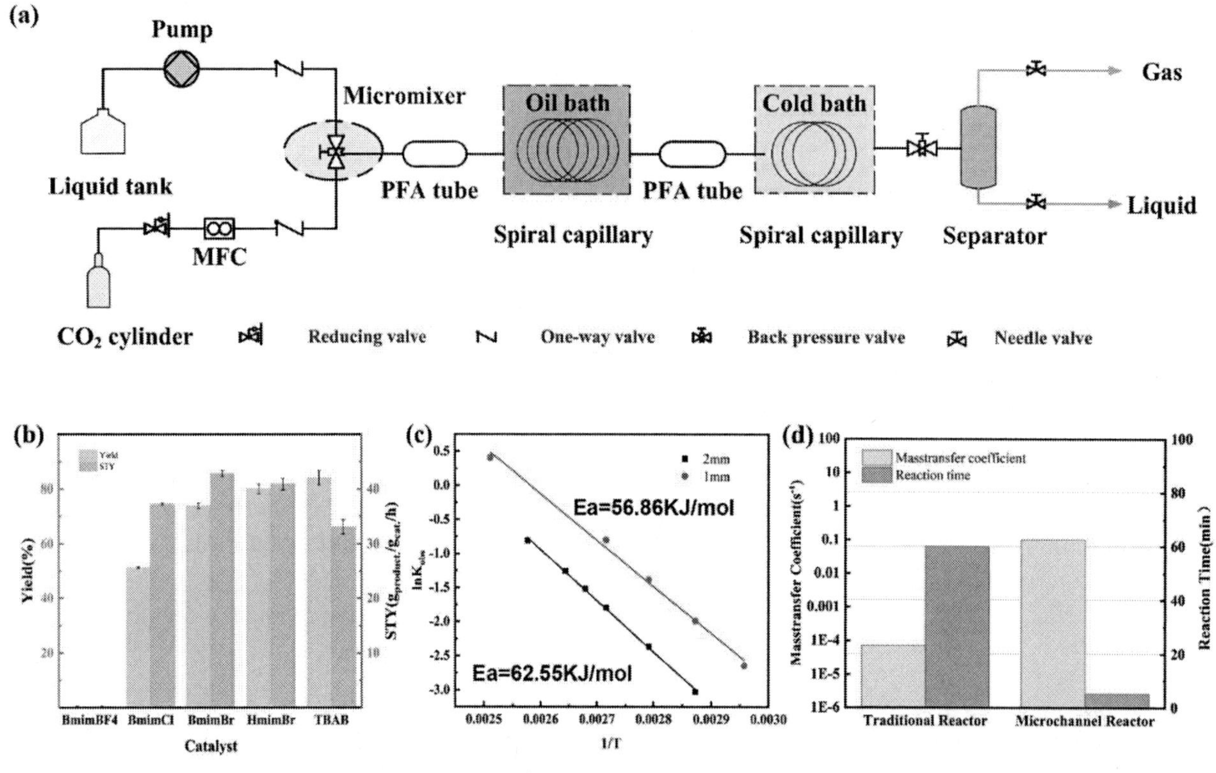

196f Structural Reorganization of Ionic Liquid Electrolytes By Charge/Discharge Circle

Kun Zhang, Guohui Zhou, Timing Fang, Xiaomin Liu

The behavior of forming layers near the electrode surface is an important topic for the energy storage with ionic liquid (IL) electrolytes [1]. It is found that surface active ionic liquids (SAILs) and nonamphiphilic ionic liquids (NAILs) form different electric double layer structures near the electrode, which leads to great differences in energy storage between the two kinds of ILs [2]. Molecular dynamics (MD) simulation can reproduce the microstructure and process, and can be used to analyze the changes of SAILs or NAILs electric double layers near the electrode. We focus on IL layers near the electrode surface of graphene by MD simulation and get some change rules of charging/discharging circle. The above findings could be used to explain more complex experimental phenomenon and serve as the guide for designing ILs-based electrolytes.

Recent experimental results showed that SAILs show better charge storage performance on charged surface than NAILs [2]. Near the positive electrode of [C_4mim][AOT] system, it was found that $[AOT]^-$ would easily enter into the interfacial layer with a V-type conformation during the charging process by MD simulation. Compared with the simulated initial state, there were more V-type anions in the electrolyte after a process of charging and discharging, which is an evidence that the electrolyte was optimized. It provides an alternative way to optimize the IL electrolyte [3]. On the other hand, near the negative electrode of [$C_n$mim][AOT] and [$C_n$mim][BF_4] systems, the structural transition of imidazolium-based NAILs and SAILs in a two-stage discharge process was discussed [4]. The two-stage discharge process of NAILs was more obviously, and it was slightly gentle for SAILs. It was revealed that the quick decrease in electric energy during the first stage of discharging was mainly due to the change of cations shapes near the negative electrode instead of their movement; the slow decrease in electric energy during the second stage of discharging was caused by overscreening and slow movement of cations from the negative electrode. Unlike the conventional view, it is concluded that the large scale movement of ions plays a relatively minor role in releasing electric energy. It will help us to understand the mechanism of energy release and provide a theoretical basis for the rational design of ILs-based electrolytes.

References

[1] Y. Wang, , H. He, , C. Wang, , Y. Lu, , K. Dong, , F. Huo, S. Zhang, Insights into Ionic Liquids: From Z-Bonds to Quasi-Liquids, *JACS Au* **2** (2022) 543-561.

[2] X. Mao, P. Brown, C. Cervinka, G. Hazell, T. A. Hatton, Self-assembled nanostructures in ionic liquids facilitate charge storage at electrified interfaces, *Nature Materials* **18** (2019) 1350-1357.

[3] K. Zhang, G. Zhou, T. Fang, K.Jiang, X. Liu, Structural Reorganization of Ionic Liquid Electrolyte by a Rapid Charge/Discharge Circle, *Journal of Physical Chemistry Letters* **12** (2021) 2273-2278.

[4] K. Zhang, G. Zhou, , T. Fang, X. Tang, X. Liu, Different shapes based on ionic liquid leading to a two-stage discharge process, *Journal of Materials Chemistry A* **10** (2022) 7684-7693.

(a)

(b)

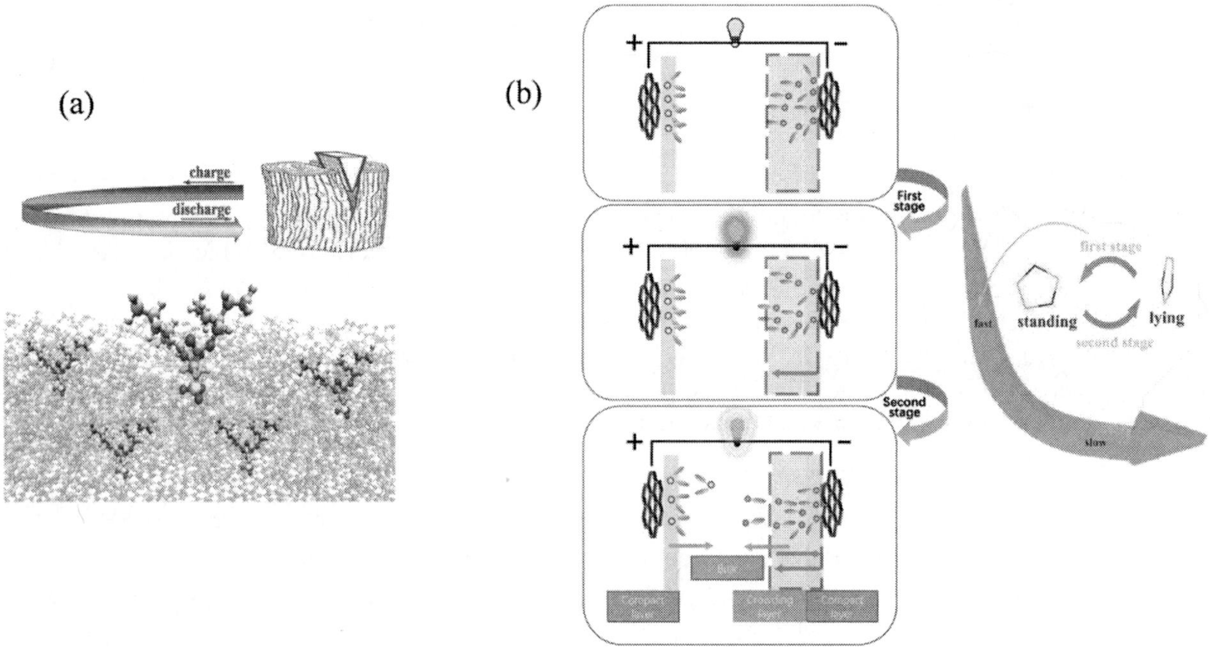

Figure 1: (a) V-type anions, and (b) the two-stage discharge process for SAIL.

196g Synthesis, Characterization, and Assembly of All Polyelectrolyte Diblock Copolymers from Poly (ionic liquid) and Weak Polyelectrolyte Blocks

Kayla Foley, Keisha Walters

Poly(ionic liquids) (PILs) are a special type of strong polyelectrolytes that combine the diverse functionality and unique properties of ionic liquids (ILs) with the mechanical stability, long-range ordering, and processability of polymers. In recent years, PILs have found application in a variety of fields including battery electrolytes, separations, nanomaterials, and more. In one growing field, PILs have been combined with highly incompatible neutral polymers to form block copolymers (BCPs). These systems have been used in forming polymer films and membranes with unique nanostructures such as lamellae, cylinders, spheroids, and bicontinuous structures. Depending on the structure, these morphologies can enhance the properties of PILs materials, such as ion conductivity in battery electrolytes or interfacial behavior in coatings. However, only a few studies have investigated block copolymers composed of PILs with weak polyelectrolytes, which can modulate their charge density based on degree of protonation. Such materials show conformational changes in response to environmental stimuli such as changes in pH and salt. Synthesizing and characterizing PIL-b-polyelectrolytes will aid in the development of novel all polyelectrolyte materials for applications in stimuli response, polymer battery electrolytes, and flexible electronics. In this work, (PIL-b-polyelectrolyte) BCPs are synthesized through the SARA ATRP controlled free radical polymerization technique. Specifically, poly[vinylbenzyl butyl imidazolium bis(trifluoromethanesulfonyl) imide] was polymerized utilizing a tertiary amine weak polyelectrolyte macroinitiator. The synthesized polymers were characterized for their chemical properties and molecular weight through GPC, NMR, and FITR. The solubility and solution behavior of these block copolymers were investigated with DLS. Thermal properties were characterized with DSC and TGA. A preliminary examination of the block copolymer self-assembled films on wafers was also performed with optical microscopy, SEM, and combined FTIR-AFM.

196i Dynamic and Structural Properties of Ionic Liquid Electrolytes

Jiahuan Tong, Feng Huo, Bilin Zhuang, Suojiang Zhang, Xiaodong Liang, Nicolas von Solms

Lithium-ion batteries are currently the most popular and widely used energy storage devices. They are almost omnipresent in modern society in portable devices, electrical vehicles, and energy storage stations. There has been a growing demand to improve the battery efficiency, durability, and sustainability. Ionic liquid electrolytes (ILE), which uses ionic liquids as the solvent, is a promising candidate for improving the safety and the cycling performance of the lithium-ion batteries.[1],[2] However, the optimal ILE composition and the charge-transfer mechanism in ILE have not been elucidated, hindering the wide adoption of ILEs in industries. In this work, using molecular dynamics simulations, we aim to present a systematic study on ILE performances, search the optimal ILE composition, and clarify the transfer mechanism of lithium ion in the ILE. We reveal that ILE has higher conductivity than conventional organic electrolytes at high lithium-salt concentrations. Furthermore, we discover that the lithium ions are transferred in the form of lithium-anion clusters in the ILE.

196j Highly Dispersed Ionic Liquids in Mesoporous Molecular Sieves Enable a Record NH_3 Absorption

Zhiyong Li

Ammonia (NH_3) has high hydrogen energy density and is a promising sustainable energy in the future. However, developing simple, stable and efficient materials for NH_3 uptake and storage remains a challenge. Here, a family of stable ionic liquid hybrid materials have been developed by simple impregnation of Zn-based ionic liquids (ZnILs) into stable mesoporous molecular sieve FDU-12 for high-efficient NH_3 capture. The ZnIL hybrid adsorbents exhibit a record NH_3 adsorption capacity (25.2 mmol/g at 25.0 °C and 1.0 bar), along with exceptionally high uptake at high temperature (15.2 mmol/g at 80.0 °C and 1.0 bar) and low NH_3 concentration (5.0 mmol/g at 25.0 °C and 10000 ppm). Notably, $[C_1NH_3][Zn_3Cl_7]$@FDU-12-40% shows almost unchanged breakthrough performance after 5 cycles, indicating excellent cycling stability of these hybrid adsorbents. It is revealed that such superior NH_3 capacity is mainly attributed to the highly dispersed ZnILs in FDU-12, which facilitates the coordination and hydrogen bonding interactions of NH_3 with the ionic liquids.

196k Rare-Earth Separation Based on the Differences of Ionic Magnetic Moment Via Quasi-Liquid Strategy

Fujian Li

Rare-earth separation based on the differences of ionic magnetic moment via quasi-liquid strategy

Fujian Li[1, 2*], Bangyu Fan[1, 2, 3], Yanyan Cheng[1, 2, 4], Zhongzhi Wang[1, 2, 4],

Suojiang Zhang[1, 2], Lu Bai[1, 2], Na Wang[1, 2] and Xiangping Zhang[1, 2*]

1 CAS Key Laboratory of Green Process and Engineering, State Key Laboratory of Multiphase Complex Systems, Beijing Key Laboratory of Ionic Liquids Clean Process, Institute of Process Engineering and Innovation Academy for Green Manufacture, Chinese Academy of Sciences, Beijing 100190, PR China

2 Ganjiang Innovation Academy, Chinese Academy of Sciences, Ganzhou 341000, PR China

3 College of Chemistry, Nanchang University, Nanchang 330031, PR China

4 School of Rare earths, University of Science and Technology of China, Ganzhou 341000, PR China

*Corresponding author, Email: fjli19@ipe.ac.cn, xpzhang@ipe.ac.cn

The separation of rare earth elements (REEs) is particularly difficult due to their similar physicochemical properties. Based on the tiny differences of ionic radius, solvent extraction has been developed as the "mass method" in industry with hundreds of stages, extremely intensive chemical consumption and large capital investments. The differences of the ionic magnetic moment among REEs are greater than that of ionic radius. Herein, a novel method based on the large ionic magnetic moment differences of REEs was proposed to promote the separation efficiency. REEs were firstly dissolved in the ionic liquid, then the ordering degree of REEs was improved with the Z-bond effect, and finally the magnetic moment differences between paramagnetic and diamagnetic REEs in quasi-liquid (QL) system were enhanced. Taking the separation of Er/Y, Ho/Y and Er/Ho as examples, the results showed that Er(III) and Ho(III) contained ionic liquids(ILs) had obvious magnetic response, while ILs containing Y(III) had no response. The separation factors of Er/Y and Ho/Y were achieved at 9.0 and 28.82, respectively. It was obviously higher than that of the traditional separation method with $\beta_{Er/Y}$ and $\beta_{Ho/Y}$ of 1.4 and 1.64 by P507. Magnetic separation via quasi-liquid system strategy provides a possibility of novel, green, and efficient method for rare earth separation.

Keywords: rare earth element, different magnetic moment, magnetic separation, ionic liquid

124a Ammonium Formate As a Safe, Energy-Dense Electrochemical Fuel Ionic Liquid

Zachary Schiffer, Sayandeep Biswas, Karthish Manthiram

Hydrogen has been investigated as an energy carrier for clean electrons; however, its low energy density and compression losses limit its use. Accordingly, solid and liquid hydrogen carriers have been proposed, including liquid ammonia, formic acid, methanol, etc. While these fuels have significant potential, many do not meet the energy efficiency requirements to outperform hydrogen. For example, liquid ammonia has a high hydrogen content (110 kg H_2/m^3) compared to compressed hydrogen gas (40 kg H_2/m^3), and it is already produced, transported, and stored at scale. Formic acid has a slightly higher hydrogen content (53 kg H_2/m^3) than compressed hydrogen gas and is also produced at scale. In both cases, there has been recent progress on production with renewable electricity. However, both chemicals suffer from safety issues because they are caustic and corrosive, as well as difficulties with efficient energy release at scale.

In this work, we investigate ammonium formate, a combination of ammonia and formic acid, as an energy carrier. Its production is as cheap and simple as its constituent molecules, and it is a solid at ambient conditions, making it safe to transport and store. We demonstrate an electrochemical cell that takes advantage of the fact that ammonium formate can form an ionic liquid at temperatures slightly above 100°C. In this cell, hydrogen evolves at the cathode with ca. 100% Faradaic efficiency and formate oxidizes to carbon dioxide at the anode with ca. 100% Faradaic efficiency. In theory, ammonia could be oxidized at the anode; however, ammonia evaporates at this operating temperature and can be oxidized using a second, modular ammonia fuel cell. Because molten ammonium formate is conductive, the electrolyte is majority fuel, generating a large thermodynamic driving force. Overall, this system represents a new class of electrochemical fuel ionic liquids and results in modular release of hydrogen with potentially zero net carbon emissions.

124b Fossil Fuel Assets and Wastewater Resources Integrated SOLID OXIDE Electrolytic Cell for a NOVEL Hydrogen Production

Lateef Jolaoso, Javad Asadi, Pejman Kazempoor

ABSTRACT

Water and energy are two inseparable resources needed for the existence, wellbeing and economic sustainability by human. The production of energy requires substantial amount of freshwater. There is an increase in energy demand due to several factors like the population growth. During the continuous expansion of an energy sector, the mix of the deployed technologies to produce fuels and generation of electricity controls the associated impact on freshwater resources in that area. This work explores the nexus model by focusing on the design, performance and economic estimation of a novel, highly efficient and environmentally benign hydrogen production process which can be integrated with several power production units, including coal, natural gas-fired power plants and renewable energy sources. The system employs flue gas from coal power plants and various wastewater sources to generate steam as the process's feedstock for the SOEC unit. The SOEC and the balance of plant components are designed, simulated, and integrated using Aspen HYSYS Software. The SOEC model is experimentally and numerically validated, and a piece-wise performance evaluation of the system is carried out. The result of this innovative system shows overall thermal-to-hydrogen efficiency of 56.3% and 61.97% if the heat of the high-temperature exiting species are put to use and the SOEC shows a voltage efficiency of 92.9% which can be improved depending on the energy source and by full energy Integration. The techno-economic analysis of the system gives a Levelized hydrogen cost of $2.99/kg, which can be upgraded to meet the U.S. Department of Energy (DOE) targets and the life cycle analysis shows using solar PV to power the system is of greater economic prospect . The study offers insights into water-energy nexus for the researchers and policymakers for decision making and technology development.

Keywords: Solid oxide electrolytic cell (SOEC), Hydrogen production, Flue gas, Water-energy nexus, techno-economic analysis (TEA)

INTRODUCTION

The production of hydrogen from renewable and sustainable sources is a key to the green hydrogen economy. Water electrolysis has gained attention and looks promising as the process is greenhouse gas-free and yield 100% pure hydrogen. Even though water is abundant, its accessibility is a great challenge, and its use for this process will undermine water security for domestic and agricultural use. One of the major crises faced in this generation due to its increasing population is the availability of clean drinking water, particularly in developing countries. Even the developed ones can only access 2% of the available water[1]. China, most especially its coal-rich region, is faced with water scarcity[2] and projected to be severe in 2050 and the arid and semi-regions like the Middle East and the western U.S. [3].

There are three major commercialized and conceptualized water electrolytic cells for the generation of hydrogen in recent times [4] which include alkaline electrolytic cell (AEC), the most matured technology [5], polymer electrolyte membrane (PEM) electrolytic cell (PEMEC) and solid oxide electrolytic cell (SOEC). The AEC challenges range from low current density to low operational pressure, which aimed to be solved with the advent of PEMEC[6]. Likewise, the PEMEC suffers from high capital cost, pure water requirement, and increased system complexity, stemming from its operational pressure [6, 7]. However, the SOEC is an evolving and promising technology that is anticipated to overcome the setbacks plaguing these two electrolyzers [8, 9]. Solid ion-conducting ceramics are used as electrolytes in the SOECs, which operate at high temperatures ranging from 650-1000 °C [9]. Several works have been done on this technology both in the laboratories [8] and industries to commercialize it [9]. Kazempoor and Braun[10] modeled and evaluated a SOEC system to serve as a primary system-level performance prediction tool. Several contemporary studies on SOEC focused on hydrogen production directly from water and steam [11-13]. For better system optimization high round-trip efficiency, SOEC has been designed and studied alongside its reverse counterpart, solid oxide fuel cell (SOFC), to form regenerative solid oxide cell (rSOC) [14-16]. To eliminate the uncertainty concerning future cost and high capital cost of this technology, recent literature evaluates its techno-economic viability to encourage investments[17-19].

This work is distinctive in its water electrolysis in that it explores the flue gas from power plant and wastewater as a feedstock into the hydrogen production unit integrated with a different energy source. The net energy and techno-economic analyses are carried out on the entire system for performance evaluation and sensitivity analysis on the system's critical parameters.

SYSTEM AND MODEL DESCRIPTION

The proposed system is shown in fig. 1. It comprises two sections, namely the upper steam generating section and the high-temperature SOEC unit- hydrogen production unit.

In the upper section, high pressurized steam is generated from the wastewater reservoir after it is pumped to the water treatment and recovery unit (WTRU) where it is evaporated at atmospheric pressure and 101 °C. This is mixed with the steam, which is concurrently generated from the power plant flue gas. The flue gas is cooled to 25°C in the condenser for moisture recovery, and the heat regained is used in the WTRU. The recovered water is immediately evaporated and combined with the steam from the WTRU. The mixed stream is sent to the pre-heater to elevate its temperature to 750-850°C before being fed into the SOEC to lower the required entropic heat for decomposition. Air from the ambient environment is compressed, preheated, and supplied to the system to flush out the produced oxygen from the electrolysis reactions in the SOEC. The exiting hydrogen product is stored in a compressed hydrogen tank.

The water decomposition reaction taking place in the SOEC is represented by the anodic oxidation and cathodic reduction reaction below

Planar model configuration was used in the SOEC design with the system comprising of the balance of plant (BoP) such as condenser, heater, compressor, mixer, evaporator, separator, pump, heat exchanger to finally actualize a working model which are thermodynamically model using hysys. All other model assumptions are similar to that of Kazempoor and Braun[10].

RESULTS AND DISCUSSION

This section summarizes some of the important results of this work. The SOEC is in the developing stage, and this novel work is evaluated and validated. The system generated 242400 kg/day of hydrogen from the flue gas from a power plant with a capacity 333.7 MW. The voltage efficiency of the SOEC system is 92.9% at 850 °C, and the overall thermal efficiency of the designed plant is 61.97%, which seems to be very prosperous compared to other plants of the same size. Even though it is shown from hysys that is potential to save over 25% of the utilized energy in the system which reduces cost of production and optimize productivity. To live up to the progressive development in different sources of energy, the model is designed to be integrated with several energy sources. The Levelized cost of hydrogen (LCOH) production with some of these energy sources is evaluated and has shown in fig. 2. The LCOH ranges from $2.99-5.96/ kg hydrogen with the solar photovoltaic (P.V.) given the cheapest cost assuming the U.S. Department of Energy (DOE) target of 3 cent/kWh by 2020 and beyond. As such, the life cycle analysis (LCA) from cradle to gate for the plant is performed in SimaPro using the solar P.V. and the coal power plant to measure the severity of the environmental impact. Obviously, the COx emission with the coal power plant is higher. However, SimaPro revealed the higher effects of some other critical parameter factors like the human health, ecosystem and particulate matter formation associated with it as shown in fig. 3.

CONCLUSIONS

The novel model designed in this work shows the possibilities of different energy integrations for hydrogen production and redefined a new approach to water-energy nexus. The LCOH of hydrogen can be as low as $2.99 per kg hydrogen, and this is a close route to meeting the DOE target for the cost of hydrogen for distribution and as potential energy storage. Powering the system with solar PV is of high economic prospect and further work needs to be embarked on to establish the exact environmental impact of large-scale hydrogen production to meet up the global requirement for sustainability and a green economy.

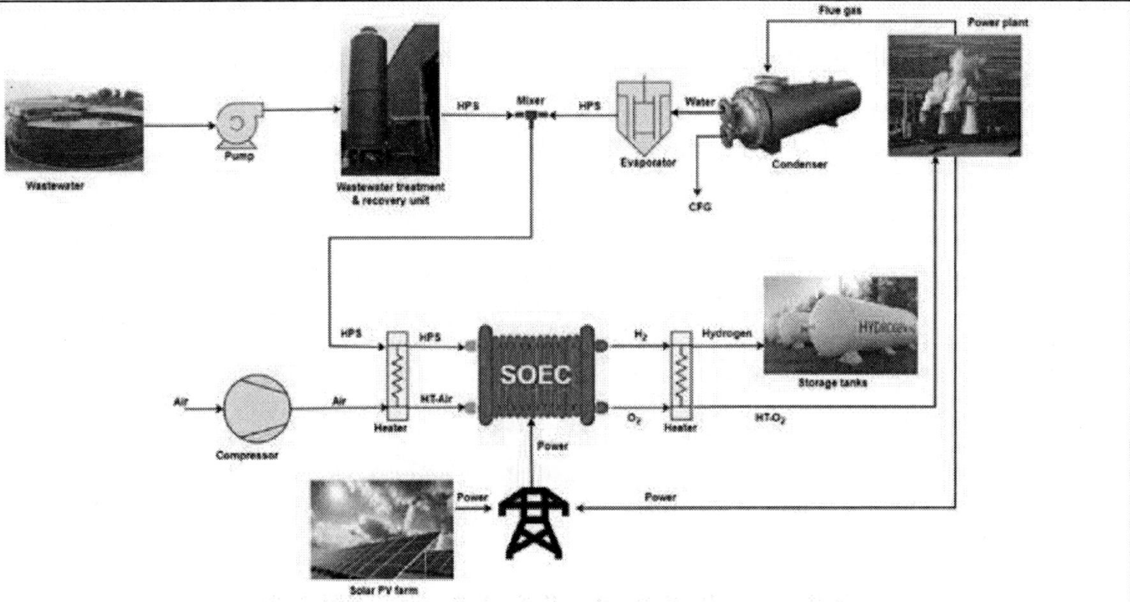

Fig 1: Water-energy integrated system for hydrogen production

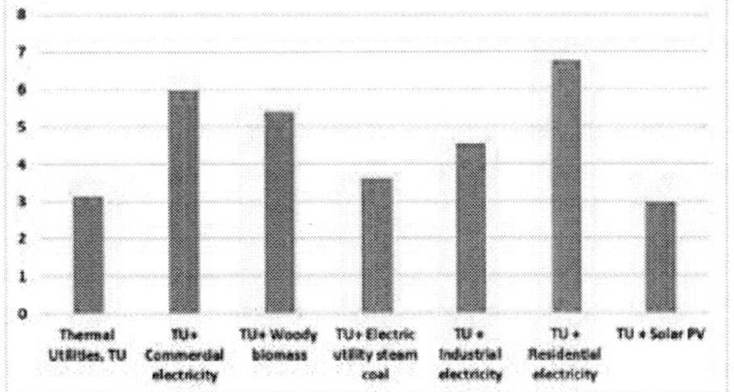

Fig. 2: Levelized cost of hydrogen (in $) with different feedstocks

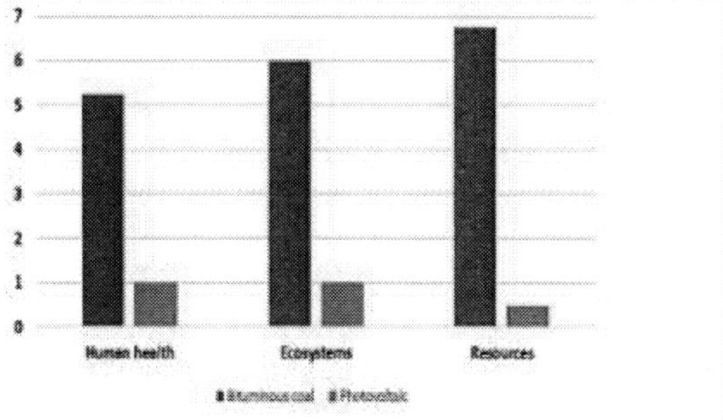

Fig 3: Damage assessment for electrolysis reaction by solar PV and coal power sources

124c Enhanced Phase Change Materials for Ocean Thermal Energy Conversion

Jian Liu, Brianna Friedman, Hyunjun Jung, Aljon Salalila, Habilou Ouro-Koura, Daniel Z. Deng

Latent heat storage through phase change materials (PCM) has attracted attention in the ocean thermal energy conversion (OTEC) area and unmanned underwater vehicles (UUVs) because it provides economic energy storage solutions with small weight and volume requirements. The goal here is to develop advanced PCM with enhanced thermal conductivity to accelerate the heat transfer rate that will lead to a higher power generation per profile for UUV using PCM technology. Multiwall carbon nanotube (MWNT) modified was used as additives to enhance the thermal conductivity of hexadecane which was selected as the PCM candidate due to its appropriate melting temperature (18 ^{o}C) and availability. The dispersion of the MWNT in hexadecane was enhanced by the modification of the long carbon chain and the thermal conductivity of the hexadecane can be increased by more than 20% by adding less than 4 vol% modified MWNT. Correlations between the thermal conductivity and onset melting point with the concentration of the added modified MWNT were investigated by a thermal conductivity analyzer and differential scanning calorimeter. Volume expansion of the PCM was also measured to understand the potential impact of the enhanced thermal conductivity on the conversion of thermal energy to hydraulic energy.

124d Substituted Zeolites As Promising O_2 Sorption Pump Materials: A Density Functional Theory Study

Steven Wilson, Ellen B. Stechel, Ivan Ermanoski, Christopher L. Muhich

Many chemical processes depend on having an environment that is low in oxygen partial pressure (P_{O2} < 100 Pa); sorption pumps are a promising route to establishing that environment by either oxygen pumping or oxygen separation from an inert gas. Near ambient sorption-based processes rely on either pressure or thermal swings, requiring no moving parts, no electricity, and neither very high nor very low temperatures. In this work, we use *ab initio* calculations to explore zeolites as a class of materials for sorption-based oxygen pumping/separation. Our calculations indicate that while the neat $AlPO_4$ composition of ALPO-5 zeolite (IZA-code AFI) does not adsorb O_2, O_2 does adsorb on zeolites selectively substituted with transition metals and metalloids and hence, can enable separation and pumping. ALPO-5 substituted with Si, Ge, Sn, Pd, Pt, Ti, V, Cr, Mn, Zr, Mo, Hf, W, Ce, and Pr provides adsorption energies ranging from -0.19 to -3.92 eV, (-) indicates exothermic process. Additionally, we provide a comprehensive understanding of what controls the adsorption energy: 1) the substitutions must be able to adopt an oxidation state that is more positive than the cation it replaces, and 2) the size of the pore into which the O_2 adsorbs to the wall. We compare these trends found in substituted ALPO-5 to three other $AlPO_4$ compositional zeolites VPI-5, SSZ-51, and ALPO-52, IZA-code VFI, SFO, and AFT respectively. We find that the factors in controlling binding energy are confirmed in these three zeolites as VPI-5, which contains on pore ~3 Å larger in diameter, substituted with similar elements results in higher binding energies than ALPO-5. Furthermore, ALPO-52 having smaller pores than ALPO-5 is less favorable to O_2 binding. Lastly SSZ-51s structure is similar to ALPO-5s, with only a 0.5Å dimeter difference in the largest pore, and results in almost identical O_2 binding behavior. Additionally, we conduct a thermodynamic analysis of a thermal swing cycle to approximate the optimal O_2 binding energy for low energy O_2 pumping/separation. We find that the minimum energy cost likely occurs when the adsorption energy is in the range of 0.75 – 1.00 eV (72 – 97 kJ·mol^{-1}), which corresponds to Ge, V, Pt, or Ce substituted ALPO-5. While also investigating the thermodynamic selectivity towards O_2 adsorption as compared to other air molecules like N_2, CO, CO_2 etc.

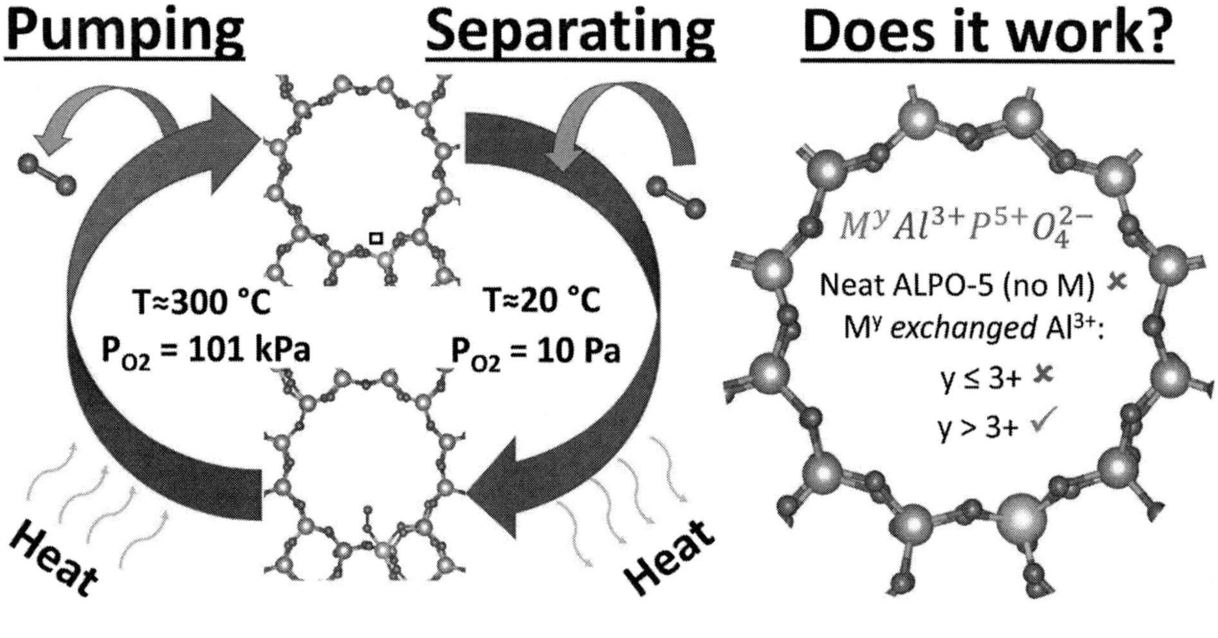

532by Investigation of the Thermochemical Reaction Pathways of Solid-State Ammonia Borane with Chemical Oxidizers

Prithwish Biswas, Yujie Wang, Hyuna Kwon, Bryan M. Wong, Michael Zachariah

Ammonia Borane (NH_3BH_3/AB), a widely explored molecule for low temperature hydrogen storage applications, is also an attractive fuel for energetic materials, provided that its constituent boron and hydrogen can be fully oxidized. Investigation on the high temperature thermochemical reaction kinetics of AB with solid state oxidizers used in energetic materials have not been performed earlier. Using in-situ time-of-flight mass spectrometry supported by ATR-FTIR, DSC, NMR, and DFT calculations, we show that the reactions of AB with oxidizers like $KClO_4$, CuO, and Bi_2O_3 are mediated by $[NH_3BH_2NH_3]^+[BH_4]^-$ (DADB) formation, resulting in its kinetic entrapment into low-energy BNH_x clusters that are resistant to further oxidation, thus limiting complete energy extraction from AB. On the contrary, we find that with NH_4ClO_4 as an oxidizer, the presence of NH_4^+ ions causes AB to follow an alternative reaction route forming $[NH_3BH_2NH_3]^+[ClO_4]^-$ instead of DADB, thus inhibiting the formation of BNH_x species, hence enabling its complete oxidation. This alternative reaction route enables the extraction of the complete energy content of AB, causing the AB/AP system to release energy ~27 times faster than the AB/$KClO_4$ and ~7 times faster than standard propellant Al/AP system.

257a *Low-Cost Waste Feedstock Conversion to Sustainable Aviation Fuel*

Karthikeyan Ramasamy, Michael R. Thorson

To decarbonize the aviation sector and to meet the 2050 net zero emission goals, developing renewable fuel technologies are imminent. Aviation sector is hard to electrify due to its current status of the technology and the required energy density, and liquid fuels will continue to play a major role for the foreseeable future. So, developing technologies to convert feedstocks that are considered waste and low values provide great opportunity and converting, underutilized, wet organic and municipal solid wastes to "drop-in" sustainable aviation fuel (SAF) represents an immediate opportunity to decarbonize the aviation sector. This presentation will discuss the current status of the hydrothermal liquefaction technology to convert wet waste feedstocks and alcohol to jet technology to convert the dry waste feedstocks to produce SAF.

257b Process Systems Engineering Contributions to Flexible Carbon Capture

Michael Baldea, Joan Brennecke, Kyeongjun Seo, Mark Stadtherr

Carbon capture remains a key component of the transition towards sustainable energy. Optimal design and operation of carbon capture plant are critically important in ensuring that this transition is economically feasible. In this presentation, we discuss recent progress in tools and algorithms for modeling, simulating and optimizing solvent-based carbon capture facilities, with a focus on processing flue gas from power plants. We expand on the multi-scale aspect of the problem, whereby the design of the solvent and of the entire plant are considered simultaneously, while also taking into account the time-varying operation of the power plant as determined by fluctuations in power demand on the grid. Several design and operational innovations that enhance the flexibility of the capture process under these circumstances will be presented, along with novel solvent regeneration strategies that minimize thermal degradation.

257c Process Engineering in Developing and Commercializing Breakthrough Technologies for Sustainability

Yizu Zhu

The issues in achieving sustainability for breakthrough product and process technologies from ideation to manufacturing present challenges to scientists and engineers. This presentation addresses the important role of integrated process engineering plays where data are sparse in Conceptual Process Design, Scaling-up, Commercial Design, Manufacturing Optimization and Process Revamping with some examples.

455a A Low-Temperature Hollow Fiber Membrane Reactor for Intensified Propane Dehydrogenation

Lu Liu, Antara Bhowmick, Dongxia Liu, Chen Zhang

Propylene is a crucial petrochemical feedstock for a number of bulk chemicals and polymers. While steam cracking remains to dominate propylene production, propane dehydrogenation (PDH) has been increasingly practiced to address the gap between propylene demand and production. The propane conversion of PDH reactions is challenged by thermodynamic limitation and reaction endothermicity. High reaction temperatures (above 550 °C) are required to obtain attractive propane conversion, which leads to rapid catalyst deactivation. Membrane reactors can potentially address this challenge by integrating reaction and separation to reduce the capital cost, energy consumption, and footprint of propane dehydrogenation.

In this talk, we will present a novel hollow fiber membrane reactor for intensified propane dehydrogenation and product separation. The membrane reactor consists of novel hollow fine fiber membranes (diameter ~ 0.3mm) for in situ hydrogen removal and platinum-based catalysts for propane activation. The hollow fine fiber membranes have high membrane packing density, which allows small membrane reactor footprint and modular propylene production. The hollow fiber membranes efficiently removed the hydrogen product from the membrane reactor and enhanced the propane conversion by 200% over the equilibrium conversion. This gave commercially attractive propane conversion (above 30%) and high propylene selectivity (above 98%) at 450 °C, which is at least 100 °C lower than commercial PDH processes. The low reaction temperature suppressed catalyst deactivation, thereby allowing the intensified membrane reactor to provide unprecedentedly stable conversion and selectivity to outperform many PDH membrane reactors reported in literature.

455b Experimental Optimization of Process Parameters for Ammonia Production

Fouzia Hasan Nowrin, Mahdi Malmali

Decarbonizing ammonia (NH_3) production is one of the major targets of net zero by 2050, as the industrial synthesis route is responsible for releasing 1.6 tonnes of CO_2 per tonne of NH_3. One pillar of decarbonization is to enhance the process energy efficiency, while drastically reducing the capital costs. A promising route is to replace phase-changing condensation in the Haber-Bosch (HB) synthesis loop with high temperature absorber columns that enable production at lower pressure. Successful demonstration of the reaction-absorption synthesis loop has been reported earlier, in which cheap, abundant metal halide salts were used to selectively absorb NH_3 almost instantly at a wide temperature at lower pressure. Although absorption of ammonia on metal halides is widely investigated for heat pumps, we still know little about their performance in HB loop. The effect of synthesis loop operating conditions on the reaction-absorption process is yet to be studied. In particular, absorption is a cyclic process and induces transient behaviour in continuous operation of reaction-absorption. Optimizing reaction and absorption to maximize the ammonia production rate and purity in the product line is the subject of this presentation.

Using a $CaCl_2$-based absorber column, we conducted lab-scale tests to optimize the reaction-absorption process parameters and further improve our understanding of RXN-ABS transient behaviour. We evaluated NH_3 production in three different modes: reaction-only, reaction-then-absorption, and reaction-absorption. Reaction-only tests provided accurate measurements of the reaction rates, whereas reaction-then-absorption tests allowed to measure the extent of NH_3 removal from synthesis gas, downstream of the reactor. The reaction and absorption conditions (pressure, temperature, and space velocity) were varied to evaluate reaction and absorption performance in these tests. Later, these optimized conditions were used to evaluate reaction-absorption performance in a transient and cyclic mode. The NH_3 release conditions and absorber cycling were further studied to achieve +95% NH_3 purity downstream of the absorber, upon regeneration.

References:

Nowrin, Fouzia Hasan, and Mahdi Malmali. 'Optimizing Reaction-Absorption Process for Lower Pressure Ammonia Production.' *ACS Sustainable Chemistry andamp; Engineering* 10.37 (2022): 12319-12328.

Hrtus, Daniel J., et al. 'Achieving+ 95% Ammonia Purity by Optimizing the Absorption and Desorption Conditions of Supported Metal Halides.' *ACS Sustainable Chemistry andamp; Engineering* 10.1 (2021): 204-212.

455c Intensification of Dispersants Production Via Thin Film Evaporation for Simultaneous Reaction and Separation

Riddhesh Patel, Nasser Al Azri, Hari Mantripragada, Glenn Cormack, Nicolas Proust, Robert M. Enick, Goetz Veser

The chemical industry is currently undergoing a paradigm shift, moving away from decades old conventional processes to highly intensified, modular, and cleaner processes. The development of these new technologies is driven by multiple factors, including the need for reducing energy consumption and hence cost to remain competitive in the market, new and emerging emissions standards to mitigate climate change, new feedstocks while moving from fossil to renewable resources, and safety considerations that scale with the physical footprint of manufacturing facility. The specialty chemicals industry in particular is to-date largely operating large batch reactors of tens to hundreds of thousands of gallons in capacity and could reap significant benefits from a transition from batch to continuous manufacturing.

The present project is investigating the production of succinimide dispersants from poly-isobutylene succinic anhydride (PIBSA) and polymeric amines. The reaction occurs via two a step mechanism, an initial amidation followed by a reversible dehydration step to give poly-isobutylene succinimide as the dispersant product. In a continuous reactor, where the byproduct water from the second step remains in the product stream, the reaction is hence equilibrium limited and requires a downstream drying step to obtain the dehydrated dispersant product at full yield. We had previously identified an agitated thin film evaporator (TFE) as well-suited for drying of the highly viscous product stream.

Based on these prior studies, we now investigated the use of a TFE as a reactive separator, i.e. using the TFE as an intensified unit in which reaction and separation occur concurrently. The TFE was first modeled by integrating our previously developed reaction kinetics with mass and energy balance equations. Appropriate correlations for heat and mass transfer were identified from the literature and verified for our set-up. This performance model confirmed the feasibility of replacing the two step reactor-evaporator system with a single TFE as an intensified reactive separator. The model predictions were then validated experimentally in a next step. Overall, our study confirms that a thin film evaporator could serve as a highly intensified, integrated reactor-separator unit for reactive processing of viscous mixtures in the presence of equilibrium-limitations.

455d Thermodynamically Tuned Redox Catalyst for Ethane to Liquid Fuels Via Chemical Looping in Simple, Distributed Reactors

Luke Neal, Leo Brody, Fanxing Li

Ethane in shale gas is difficult to transport challenges for geographically isolated natural gas producers as it is not easily liquefied, and dedicated ethane pipeline capacity is limited. This leads to frequent ethane rejection, where ethane is either injected into the natural gas pipelines or flared on-site. We proposed a modular ethane-to-liquids (M-ETL) technology enabled by chemical looping oxidative dehydrogenation of ethane (CL-ODH) to ethylene followed by oligomerization to transportable liquid fuels. In a CL-ODH scheme, a redox catalyst selectively donates its lattice oxygen to selectively convert ethane to ethylene and water. Air, or steam, is then introduced in a subsequent regeneration step to replenish the oxygen depleted catalyst. The controlled feeding of reducing, purging, and oxidizing gasses is enabled by automated valve switching.

By eliminating complex feed preheating and expensive furnaces, the net exothermicity of the oxidative dehydrogenation (ODH) of ethane facilitates a simple autothermal operation. However, many previously studied mixed metal oxide materials active to CL-ODH display an endothermic ethane ODH step with a highly exothermic regeneration step. Here, we show reactor performance results of a redox material that is exothermic on both sides of the reaction. This is achieved by selecting materials with properties tuned for thermodynamically favorable oxygen donation. Perovskites, which have an ABO_3, structure consisting of one or more A-site site and B-site metal cations, can be further doped to match the thermodynamics of the CL-ODH reactions. Such material design allows for engineering of simpler reactor schemes, with internal heat transfer, similar to the Catofin process.

We present long-term, large laboratory-scale performance results of a perovskite-based catalyst that exhibits this desirable thermodynamic behavior including 1000+ hr. of reaction testing. ASPEN-based chemical process modeling further validates the ability to eliminate heat exchange, enabled by the novel materials. Additionally, updated techno-economics informed by sensitivity analysis from the process modeling of the modular reactor system will also be presented

455e Techno-Economic Analysis of Ultrasound Technology for Ethanol-Water Mixture Separation Process

Jiwoo Ha, Hyerin Seo, Junli Liu, Hao Feng, Nikolaos Sahinidis, Jonggeol Na

The main source of CO_2 emission is thermal energy-based industrial facilities such as distillation operations which accounts for about 95% of the total separation energy consumption, but the thermodynamic efficiency of the distillation column is typically under 10%.[1] Therefore, replacing the conventional thermal separation processes while reducing CO_2 footprint using electrochemical energy is highly in interest. Since the separation of ethanol from an ethanol-water mixture by classic distillation is challenging because of its azeotrope formation,[2] other technologies are required to break the azeotrope and produce anhydrous ethanol. Conventional methods for obtaining anhydrous ethanol from 5-12 wt% ethanol-water mixture reached through biomass fermentation[3] include azeotropic distillation,[4,5] extractive distillation,[6] pressure swing adsorption,[7] and a new method which is ultrasonic separation. The distillation process uses thermal energy for phase change and requires fossil-fuel-derived entrainers such as cyclohexane. Pressure swing adsorption is operated under high pressure to break the azeotrope, but ultrasonic separation does not need additional driving force or is accompanied by a phase change and is known to have better separation efficiency.[8] In this regard, we have focused on developing nonthermal, nonequilibrium ultrasonic ethanol-water mixture separation process in pilot-scale to substitute conventional ethanol-water separation technologies.

The physical mechanism of the power ultrasonic separation is the acoustic vibration which travels through a bulk liquid ethanol-water solution and releases mist consisting of liquid droplets. Ultrasound-generated liquid droplets are enriched in ethanol with higher concentrations than in the bulk solution. Ultrasonic separation can avoid the azeotropic bottleneck of distillation and low energy efficiency due to phase change.[9] Although ultrasonic separation has shown economic and environmental potential as a new ethanol-water separation method, economic assessments of ultrasonic separation on a commercial scale remain questionable due to undefined process design (e.g. pre-distillation + ultrasound, multi-stage ultrasound) and unestablished capital cost evaluation method. Therefore, the economic feasibility evaluation of ultrasonic separation has to be performed for futural commercialization. Also, the environmental impact must be evaluated for precise comparative study with conventional processes.

Here, we develop a data-driven ultrasonic separation model and report a techno-economic analysis of the ultrasound azeotropic separation process for the ethanol-water solution to produce 99.9 wt% anhydrous ethanol. We first design the conceptual design of the conventional ethanol-water separation process including azeotropic distillation, extractive distillation, pressures swing adsorption, and the new separation method, ultrasonic separation. Then, the conceptual designs developed by the process simulator are employed to predict the total annualized cost (TAC), including capital cost and operating cost of each process. In the case of ultrasonic separation, the optimal experimental condition is determined, and an additional experiment is performed based on the optimal condition. Further analysis is performed to identify the ultrasound column diameter. Then the effect of mist collection rate and transducer cost, which are likely to change with the advancement of ultrasonic separation technology, is examined. Also, we identify the global sensitivity of mist collection rate, transducer cost, utility cost, and ultrasound column diameter to understand optimal configuration compared with conventional processes. As steam used in the reboiler of the pre-distillation column is the major cost driver, we propose an optimal process design of ultrasonic separation to reduce steam use by finding the optimal pre-distillation column specification. Finally, the environmental impact of each process is evaluated and compared. Since ultrasonic separation technology is still in the development stage, the capital cost of ultrasonic separation is about twice of that of traditional technologies, but this can be supplemeted sufficiently with technology development. The operating cost of ultrasonic separation is 35-80% of conventional methods, resulting in unit production cost of $5.79/kmol ethanol which is feasible based on the prior study.[10] This study revealed that ultrasonic separation is sufficiently competitive compared to other technologies economically and environmentally on a commercial scale. The analysis emphasizes the promise that the ultrasonic separation process significantly reduces the energy usage and environmental footprint and shows the potential of the bioethanol process based on renewable energy.

[1] U.S. Department of Energy, Office of Energy Efficiency and Renewable Energy, Industrial Technology Program. (2005). Hybrid separations/distillation technology: research opportunities for energy and emissions reduction

[2] Li, G. and P. Bai (2012). New Operation Strategy for Separation of Ethanol–Water by Extractive Distillation. *Industrial andamp; Engineering Chemistry Research* **51**(6): 2723-2729.

[3] Al-Asheh, S., et al. (2004). Separation of Ethanol–Water Mixtures Using Molecular Sieves and Biobased Adsorbents. *Chemical Engineering Research and Design* **82**(7): 855-864.

[4] Castillo, F. J. L. and G. P. Towler (1998). Influence of multicomponent mass transfer on homogeneous azeotropic distillation. *Chemical Engineering Science* **53**(5): 963-976.

[5] Bastidas, Paola andamp; Gil, Ivan andamp; Rodríguez, Gerardo. (2010). Comparison of the Main Ethanol Dehydration Technologies through Process Simulation. 20th European Symposium on Computer Aided Process Engineering: ESCAPE 20.

[6] Jaimes Figueroa, J. E., et al. (2012). Improvements on Anhydrous Ethanol Production by Extractive Distillation using Ionic Liquid as Solvent. *Procedia Engineering* **42**: 1026.

[7] Mulia-Soto, J. F. and A. Flores-Tlacuahuac (2011). Modeling, simulation and control of an internally heat integrated pressure-swing distillation process for bioethanol separation. *Computers andamp; Chemical Engineering* **35**(8): 1532-1546.

[8] Sato, M., et al. (2001). Ethanol separation from ethanol-water solution by ultrasonic atomization and its proposed mechanism based on parametric decay instability of capillary wave. *The Journal of Chemical Physics* **114**(5): 2382-2386.

[9] Kirpalani, D. M., andamp; Toll, F. (2002). Revealing the physicochemical mechanism for ultrasonic separation of alcohol–water mixtures. *The Journal of chemical physics*, **117**(8), 3874-3877.

[10] Ebrahimiaqda, E. and K. L. Ogden (2017). Simulation and Cost Analysis of Distillation and Purification Step in Production of Anhydrous Ethanol from Sweet Sorghum. *ACS Sustainable Chemistry andamp; Engineering* **5**(8): 6854-6862.

455f Integrated Direct Carbon Capture and Oxidative Dehydrogenation of Ethane with CO_2 at Isothermal Conditions over Dual-Functional Adsorbent/Catalyst Composite Monoliths

Khaled Baamran, Shane Lawson, Ali Rownaghi, Fateme Rezaei

Combining CO_2 adsorption and utilization in oxidative dehydrogenation of ethane (ODHE) into a single bed is an exciting way of converting a harmful greenhouse gas into marketable commodity chemicals while reducing energy requirements from two-bed processes. However, novel materials should be developed for this purpose because most adsorbents are incapable of capturing CO_2 at the temperatures required for ODHE reactions. Some progress has been made in this area; however, previously reported dual-functional materials (DFMs) have always been powdered-state composites and no efforts have been made toward forming these materials into practical contactors. In this study, we report the first-generation of structured DFM adsorbent/catalyst monoliths for combined CO_2 capture and ODHE utilization. Specifically, we formulated Cr_xV_x-CaO/ZSM-5 monoliths by 3D-printing inks with $CaCO_3$ (CaO precursor), insoluble metal oxides, and ZSM-5. In this study, bi-function materials (BFMs) comprising of CaO and adsorbent and bimetallic Cr_xV_x@ZSM-5 catalyst (Cr_x and V_x are the metal ratio) are synthesized and processed into 3D-structures (adsorbent-catalyst monoliths) using additive manufacturing (3D-printing technique) method. The obtained DFM adsorbent/catalyst monoliths were characterized by XPS, H_2-TPR, NH_3-TPD, TGA, BET, and EDX mapping and assessed for CO_2 capture-conversion in ODHE process. Among the developed DFMs, Cr_1V_3@ZSM-5/CaO revealed the best balance between CO_2 capture (0.74 mmol/g), C_2H_4 selectivity (42%), and syngas selectivity (45%) at 650 °C due to the bimetallic incorporation that resulted in favorable redox properties for ethane dehydrogenation/reforming to ethylene and syngas. It was also found that varying the bimetallic ratio could be utilized to control the CO_2 capture, CO_2/C_2H_6 conversion, and C_2H_4/syngas selectivity. This work establishes a promising proof of concept for structuring composite materials using a facile means such as 3D-printing and expands dual function materials to the formerly not investigated application of ODHE.

References:

1. Baamran et al., *Chem. Eng. J.*, **2022**, 435, 134706.
2. Lawson et al., *Appl. Catal. B*: Env **2022**, 303, 120907.
3. Lawson et al., *Chemical Reviews,* **2021**, 121, 6246-6291.

503a Process Modeling and Simulation of Electrochemical Processes

Sadia Saberin, Chau-Chyun Chen

Numerous electrochemical technologies are being developed for energy and sustainability. Examples include electrochemical processes for chemical synthesis, batteries for power storage, fuel cells for power generation, electric vehicles, semiconductors, waste treatment, water purification, biosensors, etc. With higher efficiency and less environmental footprint, electrochemical technologies garner heightened interest among academia and industries to develop and optimize these processes. Yet, the knowledge of electrochemical fundamentals is limited to specialists and excluded from the model library of chemical process simulators readily accessible to millions of chemical engineers [Ref 1]. To accelerate the application of electrochemical technologies in chemical industry, it is important to capture the knowledge of electrochemical fundamentals in process simulators to encourage chemical engineers to practice electrochemical engineering and to expand the scope of this promising field.

This paper presents a concise summary of electrochemical fundamentals essential to model and simulate electrochemical processes. The electrochemical fundamentals include basic concepts in electrochemistry, transport phenomena, and thermodynamics. The paper also covers a review of state-of-the-art in modeling electrochemical processes and presents an example in modeling and simulating a commercially important electrochemical process to produce adiponitrile from acrylonitrile. Furthermore, this paper suggests development of specific model library modules in process simulators for electrochemical engineering.

References

1. C.-C. Chen, M.B. Gorensek, J.F. Scott, I. Palou-Rivera, "Challenges and Opportunities in Process Modeling of Intensified Chemical Processes," Chemical Engineering Progress, 2020, 116, 35-42.

503b Production of Dimethyl Ether from Coal Using Chemical Looping: Process Simulation and Exergy Analysis

Fanhe Kong, Qiaochu Zhang, Pinak Mohapatra, Andrew Tong, Liang-Shih Fan

Coal is an essential primary energy source in modern industry. An important pathway of coal utilization is to gasify coal into syngas, which is then used for downstream production of numerous chemicals such as methanol and liquid fuels. Despite its wide applications, conventional coal gasification still faces challenges such as the use of a costly air separation unit, and limited syngas H2:CO ratio. Chemical looping gasification technology is a promising alternative to replace conventional coal gasification with increased thermal efficiency and higher level of process intensification. This study presents the process analysis of chemical looping coal gasification and its integration with downstream production of dimethyl ether (DME) to reveal the advantages of the chemical looping scheme compared to conventional gasification schemes. This study discusses the simulation strategies to maintain adiabatic operation in the chemical looping system. The operating parameters of the chemical looping system are adjusted to achieve the maximum cold gas efficiency in one case study while the operating parameters of the chemical looping system are adjusted to achieve the maximum effective thermal efficiency in another case study. These two chemical looping cases involve the integration of the chemical looping system with the DME synthesis in a process that considers overall process material and heat flows and power consumption and generation. An exergy analysis is also conducted for these two integrated chemical looping cases, revealing that they both can achieve around 10% increase in exergy efficiency over the conventional gasification system.

503c A Study on Capacity Expansion Scenarios of Intensified Ethylene Oxide Process

Chinmoy Basak Mukta, Selen Cremaschi, Mario Eden, Bruce J. Tatarchuk, Paul Dimick

There is significant ongoing research on process intensification, which has the potential to offer large reductions in energy and/or feedstock use in various chemical processes [1]. Reactors are integral unit operations in chemical conversion processes and their intensification provides significant opportunities for improving the overall process performance in terms of economics, safety, and environmental impact, however adequate downstream purification and feed pre-processing are vital to realizing those advantages. So in addition to reactor intensification, there may be a need for significant modification of the rest of the process, depending on the overall intensification objective.

Ethylene oxide (EO) is a large-volume chemical raw material used in a wide range of commodity products. It is produced through a highly exothermic partial oxidation reaction. Incidentally, ethylene oxide production is one of the most energy-intensive and inefficient processes in the chemical process industry [2]. Microfibrous Entrapped Catalyst (MFEC) is a highly conductive catalyst support that has been proven to effectively intensify reactors through conductive rather than convective heat transport and large surface area [3-5]. Significant reactor intensification is possible while controlling the ignition conditions inside the reactor in the conventional ethylene oxide process [6].

In this study, we assess the impact of reactor intensification on various capacity expansion scenarios such as greenfield, brownfield, and retrofitting of the downstream separation process for ethylene oxide production. We previously reported MFEC based reactors can have higher per-pass conversion resulting in a 19% reduction in overall production cost while keeping the reactor specification the same as the conventional process [6]. Further enhancement of the MFEC reactor has been achieved through the reduction of inerts, which reduced the reactor cost and dramatically increased the capacity of the process plant. We compared different capacity expansion ranges and corresponding expansion scenarios. Our results reveal that process retrofitting is a favorable option in medium expansion scenarios and greenfield construction in large expansion scenarios.

References

[1] Tian, Y., Demirel, S. E., Hasan, M. F., andamp; Pistikopoulos, E. N. (2018). An overview of process systems engineering approaches for process intensification: State of the art. Chemical Engineering and Processing-Process Intensification, 133, 160-210.

[2] Brueske, S., Kramer, C., andamp; Fisher, A. (2015). Bandwidth Study on Energy Use and Potential Energy Saving Opportunities in US Chemical Manufacturing (No. DOE/EE-1229). Energetics.

[3] Cheng, X., Yang, H., andamp; Tatarchuk, B. J. (2016). Microfibrous entrapped hybrid iron-based catalysts for Fischer–Tropsch synthesis. Catalysis Today, 273, 62-71.

[4] Choudhury, H. A., Cheng, X., Afzal, S., Prakash, A. V., Tatarchuk, B. J., andamp; Elbashir, N. O. (2020). Understanding the deactivation process of a microfibrous entrapped cobalt catalyst in supercritical fluid Fischer-Tropsch synthesis. Catalysis Today, 343, 112-124.

[5] Sheng, M., Gonzalez, C. F., Yantz Jr, W. R., Cahela, D. R., Yang, H., Harris, D. R., andamp; Tatarchuk, B. J. (2013). Micro scale heat transfer comparison between packed beds and microfibrous

entrapped catalysts. Engineering Applications of Computational Fluid Mechanics, 7(4), 471-485.

[6] Mukta, C.B., Cremaschi, S., Eden, M. R., Tatarchuk, B. J. (2022). Techno-Economic Study of Intensified Ethylene Oxide Production Using High Thermal Conductivity Microfibrous Entrapped Catalyst. Computer Aided Chemical Engineering (Vol. 51, pp.1 Elsevier.

503d Design of Intensified Reactor for Lean Methane Emissions Treatment

Adrian Irhamna, George M. Bollas

A recent study of the Permian Basin [1] reported 194 metric tones of methane emissions to be released every hour. This makes up the methane emitted in the area to be 9.4 % of the total gas production [1], which is more than 6 times of the recommended value by the EPA [2]. Given the high global warming potential (GWP) of methane, solutions are urgently needed to reduce the impact of fossil fuels extraction and conversion on climate change [3]. Besides fossil fuels, the livestock sector accounts for about 31% of methane emissions in the US [4]. These emissions are commonly lean, diluted, and fluctuating in flow rate and composition, which makes it challenging to be converted into CO_2 in an economically sustainable manner. In this presentation, the design of a fixed bed reactor will be communicated, with the specific objective of handling lean methane emissions. The reactor is loaded with a metal-oxygen carrier and operates on a chemical looping scheme of alternating reduction-oxidation (redox) stages, while handling a continuous gas stream of lean methane. During the reduction stage, the metal oxygen carrier is reduced by the incoming gas. The oxygen carrier is, subsequently, oxidized with air in the following oxidation stage. The heat source for the overall process is provided by the exothermic reaction of methane and oxygen carrier oxidation during the oxidation stage. The metal-oxygen carrier functions as a thermal buffer that maintains the bed temperature during the alternating redox process. The feasibility of this concept is explored through dynamic simulation of a conceptual reactor loaded with a Ni-based oxygen carrier, for which reduction and oxidation kinetics have been studied extensively. The reactor model accounts for the mass, energy, and momentum balance of the fluid and solid phases inside the reactor, and the detailed kinetics of the gas, gas-solid and Ni-catalyzed reactions. We demonstrate the operational benefits of the alternating redox process by elucidating the benefits of a high-temperature zone inside the reactor bed that serves as the reaction front for lean methane conversion reactions. The resulting reduced Ni is continuously regenerated with air. We benchmark this concept against the performance of a reactor loaded with NiO operating as a reducer only, where the NiO bed gets saturated with Ni and then only promotes catalytic reforming of the lean methane to syngas. At cyclic steady-state the redox reactor system converts lean methane at 99 % conversion efficiency at a mean bed temperature of 720 °C, without the need for added heat. The proposed reactor concept was optimized to minimize methane emissions, by manipulating the air feed temperature, Ni oxygen carrier loading, and the intervals of the redox process stages. At the optimal operating condition, the reactor reached complete methane conversion at bed temperature of 850 °C, with a reaction front at the first 20% of the reactor length. The presentation will present the mathematical formulations of the reactor model and the optimization problem, simulation results that illustrate the fundamental benefits of the proposed concept and a discussion of practical limitations of the proposed concept.

Acknowledgements

This study was supported by the UTC Institute for Advanced Systems Engineering (UTC-IASE) at the University of Connecticut (UConn).

References

[1] Y. Chen *et al.*, "Quantifying Regional Methane Emissions in the New Mexico Permian Basin with a Comprehensive Aerial Survey," *Environ. Sci. Technol.*, 2022, doi: 10.1021/acs.est.1c06458.

[2] M. R. Harrison, T. M. Shires, J. K. Wessels, and R. M. Cowgill, "Methane Emissions from the

Natural Gas Industry, Volume 1: Executive Summary," *Epa*, no. June, pp. 1–24, 1996.

[3] EPA, "Overview of Greenhouse Gas." https://www.epa.gov/ghgemissions/overview-greenhouse-gases#methane.

[4] H. Aguirre-Villegas, R. A. Larson, and M. D. Ruark, "Methane Emissons from Dairy Cattle," 2016. Accessed: Apr. 04, 2020. [Online]. Available: http://www.sustainabledairy.org/publications/Documents/DairyCap_Methane_FactSheet_Final.pdf.

503e Dynamic Modeling and Explicit/Multi-Parametric Model Predictive Control Optimization of an Intensified Fluidized Bed Membrane Reactor for Oxidative Coupling of Methane

Moustafa Ali, Yuhe Tian, Shauvik De, Alexander P. van Bavel, C. Doga Demirhan, Efstratios N. Pistikopoulos

The catalytic oxidative coupling of methane (OCM) process has received intense interests from the reaction and process engineering community during the past decades. OCM provides the potential to directly convert natural gas to value-added chemicals – with reduced cost, energy consumption, and carbon emissions (Tiemersma et al., 2012; Onoja et al., 2019). However, major challenges such as low yield, fast catalyst deactivation, and reactor scaling up still obstruct the commercialization of this process. A promising solution to address these challenges is to develop innovative OCM reactor designs leveraging the recent advances in modular process intensification, e.g. micro-channel reactors and membrane reactors (Serres et al., 2012; Patcharavorachot et al., 2014). Moreover, the investigation of OCM reaction systems has mostly been focused on steady state conceptual design, while ignoring the process dynamics and control analysis to ensure actual operation under disturbances.

This work aims to develop an integrated design and control optimization approach for oxidative coupling of methane processes. An intensified fluidized bed membrane reactor (FBMR) is of interest, which has been demonstrated in our ongoing work to achieve higher C2+ yields, selectivity, and methane conversion than conventional packed or fluidized bed reactors – by keeping an low oxygen partial pressure enabled by the membrane (Ali et al., 2022). The optimal design and operational policy for the FBMR will be developed using the PAROC (PARametric Optimization and Control) framework (Pistikopoulos et al., 2015; Pappas et al., 2021). The PAROC framework consists of five steps: i) High fidelity process modeling, ii) Model approximation, iii) Formulation of the model predictive control (MPC) problem, iv) Solution of explicit control laws through multi-parametric programming, and v) Dynamic optimization for simultaneous design and control optimization. This framework has already been successfully applied to numerous conventional and intensified processes (Pistikopoulos andamp; Tian, 2022). More specifically, the high fidelity dynamic FBMR model to describe the production of ethylene and ethane from methane is developed using gPROMS ModelBuilder®.The resulting partial differential algebraic equations take into account mass balances, hydrodynamics, catalyst solid distribution, etc. A detailed 10-step reaction kinetic model is adopted from Cruellas et al. (2020). Nevertheless, given the large-scale highly nonlinear nature of the high fidelity model, a linearized surrogate model is built to balance the computational complexity for model-based control execution and the model accuracy for predictive optimization. Then, the multi-parametric model predictive control (mp-MPC) problem is solved using the Matlab® POP toolbox (Pistikopoulos et al., 2020), allowing to analytically derive the optimal explicit control policy as an affine function of state variables, disturbances, design variables, etc. The design variables considered in this work includes reactor size, temperature, catalyst particle velocity, and membrane tube design, etc. Closed-loop validation is performed to test the mp-MPC controller against the original high fidelity FBMR model, aiming to regulate the desired ethylene product purity. Via a final mixed-integer dynamic optimization formulation, the optimal FBMR design configuration and control policies can be simultaneously obtained resulting in enhanced C2+ yield, selectivity, and CH_4 conversion with guaranteed operational feasibility under disturbances.

References

Ali, M., Tian, Y., De, S., Van Bavel, A. P., Demirhan, C. D., andamp; Pistikopoulos E. N. (2022). Process Modeling, Design, and Intensification of Oxidative Coupling of Methane Process: A Comparative Study on Membrane Reactor and Fluidized Bed Reactor. ACS Spring Meeting; 2022.

Cruellas, A., Melchiori, T., Gallucci, F., andamp; van Sint Annaland, M. (2020). Oxidative Coupling of Methane: A Comparison of Different Reactor Configurations. Energy Technology, 8(8), 1–15. https://doi.org/10.1002/ente.201900148

Pappas, I., Kenefake, D., Burnak, B., Avraamidou, S., Ganesh, H. S., Katz, J., Diangelakis, N. A., andamp; Pistikopoulos, E. N. (2021). Multiparametric Programming in Process Systems Engineering: Recent Developments and Path Forward. Frontiers in Chemical Engineering, 2(January), 1–15. https://doi.org/10.3389/fceng.2020.620168

Patcharavorachot, Y., Tiraset, S., Wiyaratn, W., Assabumrungrat, S., andamp; Arpornwichanop, A. (2014). Using a membrane reactor for the oxidative coupling of methane: Simulation and optimization. Clean Technologies and Environmental Policy, 16(7), 1295–1306. https://doi.org/10.1007/s10098- 014-0813-9

Pistikopoulos, E. N., Diangelakis, N. A., Oberdieck, R., Papathanasiou, M. M., Nascu, I., andamp; Sun, M. (2015). PAROC – An integrated framework and software platform for the optimisation and advanced model-based control of process systems. Chemical Engineering Science, 136, 115–138. https://doi.org/10.1016/J.CES.2015.02.030

Pistikopoulos, Diangelakis, Oberdieck Pistikopoulos, E. N., andamp; Diangelakis, N. A., Oberdieck, R. (2020). Multi-Parametric Optimization and Control. Vol.1.

Pistikopoulos, E. N., andamp; Tian, Y. (2022). Synthesis and Operability Strategies for Computer-Aided Modular Process Intensification. Vol.1.

Serres, T., Dreibine, L., andamp; Schuurman, Y. (2012). Synthesis of enamel-protected catalysts for microchannel reactors: Application to methane oxidative coupling. Chemical Engineering Journal, 213, 31–40. https://doi.org/10.1016/j.cej.2012.09.061

Tian, Y., Pappas, I., Burnak, B., Katz, J., andamp; Pistikopoulos, E. N. (2021). Simultaneous design andamp; control of a reactive distillation system – A parametric optimization andamp; control approach. Chemical Engineering Science, 230, 116232. https://doi.org/10.1016/j.ces.2020.116232

Tiemersma, T. P., Chaudhari, A. S., Gallucci, F., Kuipers, J. A. M., andamp; van Sint Annaland, M. (2012). Integrated autothermal oxidative coupling and steam reforming of methane. Part 2: Development of a packed bed membrane reactor with a dual function catalyst. Chemical Engineering Science, 82, 232–245. https://doi.org/10.1016/j.ces.2012.07.047

503f Modeling Water-Alcohol (C1-C3) Mixed-Solvent Electrolyte Solutions with Association Electrolyte Nonrandom Two-Liquid Activity Coefficient Model

Cheng-Ju Hsieh, Chau-Chyun Chen

Recent studies have shown the utility of the association electrolyte nonrandom two-liquid activity coefficient model (association model) in thermodynamic modeling of aqueous electrolyte solutions with high surface charge density ions (e.g., Li+, rare earth element, etc.). Capturing both the self- and cross-association behavior with a species-based association term added to the original eNRTL model, the association model should be particularly effective in describing thermodynamic properties of water-alcohol mixed-solvent electrolyte solutions where, in addition to hydration of ions and formation of ion pairs, solvents self- and cross-associate. In this work, we report thermodynamic modeling of various thermodynamic and phase equilibria data with the association model for the mixed-solvent electrolyte solutions of water-methanol, water-ethanol, water-1-propanol, and water-2-propanol with alkali chlorides. The association model shows major improvement over the eNRTL model in both correlating thermodynamic properties of single-solvent electrolyte solutions and predictions for mixed-solvent electrolyte solutions.

AUTHOR INDEX